卡地亚珍宝艺术

CARTIER TREASURES
King of Jewellers, Jewellers to Kings

故宫博物院

故宫博物院
THE PALACE MUSEUM

卡地亚　　　*Cartier*

卡地亚珍宝艺术

CARTIER TREASURES
King of Jewellers, Jewellers to Kings

故宫博物院 编

Edited by the Palace Museum

紫禁城出版社
The Forbidden City Publishing House

总 目

CONTENTS

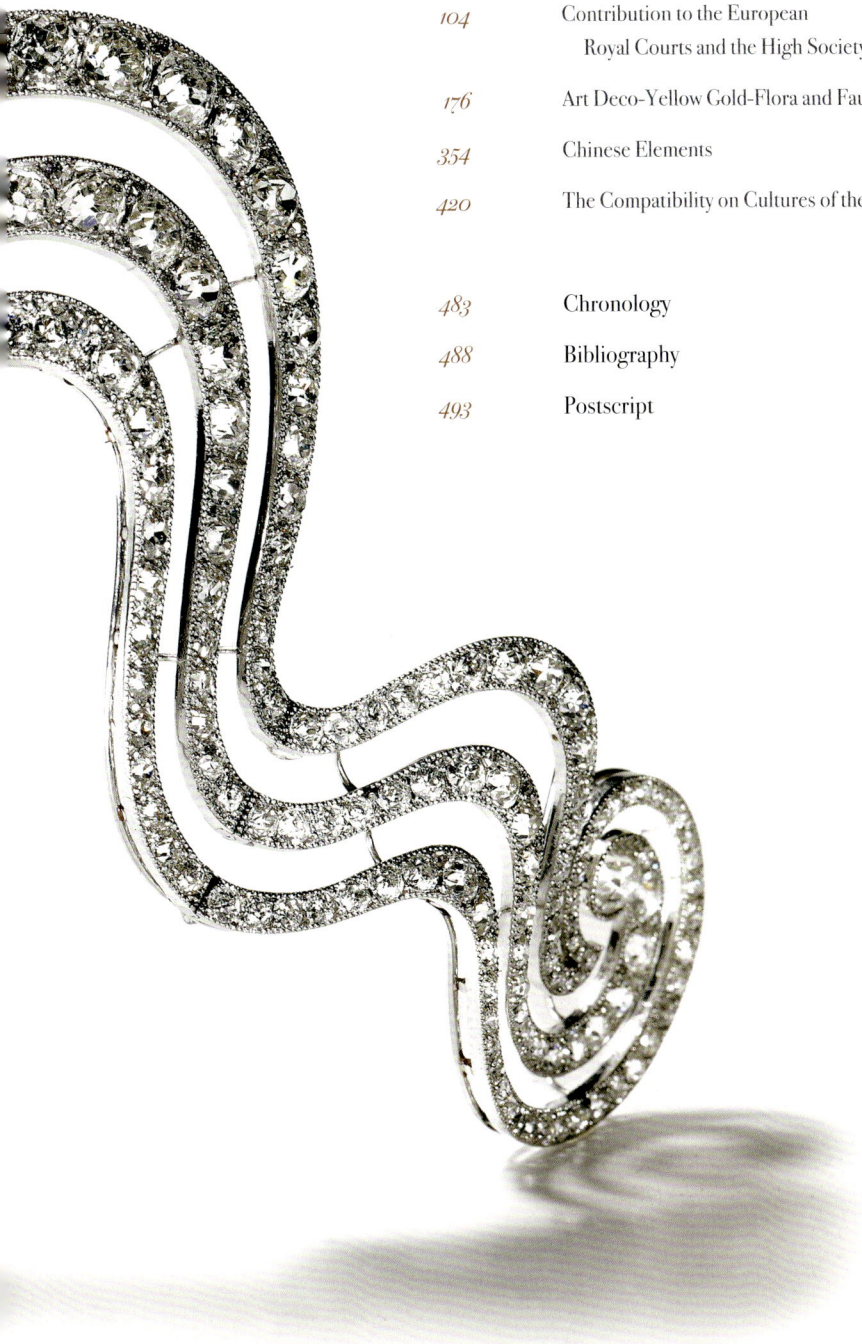

祝　辞

　　法国巴黎是世界时尚之城，旺多姆广场更是一个神奇的地方，一直吸引着热爱稀世之美的各国人士。160 多年来，卡地亚在这个以优雅与购物闻名的花都中心创造着奇迹和辉煌。

　　1847 年，才华横溢的年轻珠宝设计师路易·弗朗索瓦·卡地亚以"珠宝、饰品、时尚与新品工作坊"的名称创立卡地亚品牌，几年间便以时尚设计和精湛、细腻的工艺博得享誉国际的声誉。当拿破仑三世美丽的妻子欧珍妮在 1859 年成为卡地亚第一位王室客户后，卡地亚就与各国王室结下不解之缘，成为希腊、英国、西班牙、俄罗斯、罗马尼亚、塞尔维亚、葡萄牙、比利时、意大利、摩纳哥，甚至埃及、印度等众多王室、宫廷贵族的御用珠宝商，卡地亚也因此博得英国国王爱德华七世"皇帝的珠宝商，珠宝商的皇帝"的赞誉。

　　在卡地亚创业过程中，始终坚持借鉴世界不同民族艺术精粹的理念，坚守不断创新的本质，赋予作品广泛而深刻的文化内涵。卡地亚是云游四海的旅者、是珠宝艺术的探险家，一代代才华横溢的设计师以简洁明朗的设计风格，以铂金花环、猎豹风情、装饰艺术、异国情调、三环珠宝、奇花异兽、中国元素等主题，以及五彩缤纷的宝石与贵金属完美的结合，登峰造极的镶嵌技术，诠释着美丽、优雅和高贵，建立起卡地亚品牌特色与名望，并引领世界时尚潮流。随后，卡地亚不断拓展时尚领域，从珠宝饰品、钟表到香水、眼镜，将同样融合多种文明的法兰西优雅风情与生活艺术播撒到世界各地。并且不遗余力地赞助世界各地的文化与公益活动，扮演着亲善大使的角色。

　　1984 年，卡地亚当代艺术基金会诞生，以让更多的民众接触、认识当代艺术为宗旨，举办各种艺术展览，鼓励艺术创作，成为法国赞助艺术活动的活跃机构之一。2006 年，卡地亚启动"宣爱日"活动，此后每年的 6 月，卡地亚以著名的 LOVE 慈善手环为主角，高唱爱的颂歌，祈愿世界充满和平友爱。

卡地亚品牌已经遍及世界各地，而且在瑞士日内瓦建立了卡地亚艺术典藏室，在法国巴黎建立的资料馆，收藏了超过1 300件出自卡地亚的古董珠宝、钟表精品和数量可观的设计手稿、模具资料。经过故宫博物院与卡地亚公司的精心组织与筹备，今天"卡地亚珍宝艺术展"终于在午门展厅与广大观众见面。通过展览，我们不仅能够领略坚持创新，融汇五洲艺术风格的设计理念，追求精湛完美工艺的至高境界，而且还欣喜地看到中国文化元素对卡地亚作品的浸润和影响，从中读出卡地亚对中国文化的兴趣和诠释。为此，我们感到既亲切又自豪。穿越百年历史的卡地亚时尚精品所散发出的优雅、高贵气质和永恒魅力，将构成一场视觉文化盛宴，相信一定会带给观众审美的愉悦与享受。

感谢卡地亚和故宫博物院相关人员共同付出的巨大努力。

祝愿本次展览取得圆满成功！

故宫博物院院长

2009年9月

PREFACE

Paris is the Couture capital of the world, and Place Vendôme a landmark for beauty and excellence, admired the world over. It is here that Cartier has been working its magic and producing splendours for over 160 years.

In 1847, a young and gifted jeweller by the name of Louis-François Cartier established *La Maison Cartier*, a "manufacturer of jewellery, bijouterie, fancy decorations and novelties." The company's reputation for outstanding designs, exquisite style and intricate craftsmanship quickly spread. Empress Eugénie, consort of Napoleon III, became Cartier's first imperial patron, following which the jeweller forged an indelible bond with royal households around the world, in Greece, England, Spain, Russia, Romania, Serbia, Portugal, Belgium, Monaco, even Egypt and India. It is these royal associations that prompted King Edward VII of England to dub Cartier, "Jeweller of Kings, King of Jewellers."

By absorbing the artistic influences of other civilisations and through ceaseless innovation, Cartier endows its creations with a deep cultural relevance. This tireless explorer of the jeweller's art gleans inspiration from around the globe. Through vibrant themes such as platinum, the Garland style, Art Deco, the panther, exoticism, flora and fauna, and Chinese motifs; through unprecedented combinations of coloured gems matched with precious metals, and through superlative mounts and settings, generations of jewellers have contributed to Cartier's reputation. Cartier transcends fashion to express beauty, elegance and majesty in a unique style. Cartier has also turned its talent to other areas, extending beyond jewellery and watches to embrace leathergoods, accessories, perfumes and eyewear. It has spread French elegance around the world through its patronage of cultural and fundraising events.

Since 1984, the Fondation Cartier pour l'Art Contemporain has staged numerous exhibitions and fostered artistic creation so that more people might engage with contemporary art. The Foundation is recognised as one of the most active institutions in the sponsorship of artistic activities in France. Cartier also devotes itself to fundraising with the launch of the LOVEDAY global campaign in 2006. A LOVE CHARITY bracelet explores the true meaning of love and inspires generosity among people all over the world.

Today, the Cartier name also resonates in association with the Cartier Collection - based in Geneva, Switzerland, and Paris, France - of over 1,300 Cartier creations, jewellery, clocks and watches, together with an important collection of original drawings and plaster casts. Thanks to extensive and careful preparation by the Palace Museum and Cartier, *Cartier Treasures - King of Jewellers, Jewellers to Kings* can at last grace the Exhibition Hall of the Meridian Gate. This exhibition allows us to appreciate the innovation and multiple artistic influences that are a part of Cartier's endless pursuit of perfection. It also shows Cartier's passion for and expression of Chinese culture through the presence and influence of elements of Chinese culture in its creations. As such, we are particularly pleased and proud to welcome this exhibition, a visual and cultural feast from a jeweller which throughout its history has epitomised elegance, refinement and eternal charm. We are confident this exhibition will bring enormous pleasure and enjoyment to all its visitors.

Director of the Palace Museum

September, 2009

祝　辞

中国正成为卡地亚在全球最重要的合作伙伴。这一紧密无间的关系，源于这个泱泱大国长久焕发的迷人活力，以及卡地亚对这个辽阔国度的恒久热爱。从最初历史与风格的交汇，到如今不仅成为贸易与友谊的源泉，更是文化与艺术的纽带。

此次于故宫博物院举行的展览，堪称硕果累累。它凝聚着卡地亚与其亲密的北京合作伙伴之间的不凡友谊。同时，也代表着故宫博物院——这座世界上最负盛名的博物馆——对卡地亚作品所传承的历史文化价值的高度认可。

对此，我个人也深感欣慰。因为我本人一直为辉煌的中国文明及友好的人民所深深吸引。我经常访问中国，大约每年四次，每次都能在这片土地上发现新的机遇，邂逅新的人与事。我曾经去过最边远的中国城镇，甚至尝试学习博大精深的中文。

正是因为卡地亚与中国之间的相互交流与影响由来已久，因此随着历史的发展而愈益深刻。

早在 1888 年，卡地亚档案馆中就记录了为波内曼子爵夫人订制的一批融入中国元素的漆器。包括古塞尔街著名华人古董商店卢芹斋在内的巴黎古董商们，也曾为卡地亚提供众多古董漆器。中国神奇的动物形象、颜色及符号象征很早便被卡地亚引入其珠宝设计之中。20 世纪 20 年代卡地亚出品的很多化妆盒、首饰盒乃至小钟表的表面，都覆盖着此类闪烁迷人光泽的嵌片，令人联想到遥远的东方国度；珊瑚和玉则构成了对比鲜明的色彩组合；珠宝、魅幻时钟等作品更是大胆地赋予了麒麟、龙、佛犬、凤凰等东方神秘动物新的生命；并且路易·卡地亚本人就是这些作品的收藏家，象形文字、阴阳符号、塔顶或神祇，这些极富中国特色的图腾被卡地亚巧妙地运用于作品之中，幻化为高贵的护身符。中国美学元素令卡地亚为之着迷，直至今日，除了猎豹，龙依然是卡地亚最为推崇的动物形象之一。

皇帝的珠宝商，珠宝商的皇帝。卡地亚，这位皇室宫廷的座上宾，今天跨入故宫的恢宏门庭，置身于中国浩瀚的皇家殿阁之中，与东方皇室珍宝的呢喃共鸣。此次展览所处的午门展厅幸蒙故宫博物院院长亲自选定，位于南门中轴线上，正对故宫入口，可一览每日数千人参观这座古老皇城的盛况。能够在中国故宫举办此次回顾卡地亚艺术历史的展览，我们倍感荣幸。在此，我谨代表卡地亚对故宫博物院常务副院长李季先生、外事处处长马海轩女士、展览部副主任马继革先生、策展人宋海洋先生致以特别的谢意。感谢他们为这个宏大项目的贡献。

此次展览无疑将成为西方与东方、巴黎与北京，卡地亚与我们的中国友人在文化交流史上的里程碑。

卡地亚全球总裁兼首席执行官

伯纳德·福纳斯

PREFACE

China is on its way to becoming Cartier's foremost partner in the world. Such an exceptional link has grown from the fascinating energy that drives this vast country, and the interest that Cartier has always shown towards it. What began as historic and stylistic ties became a source of trade and friendship, as well as a cultural and artistic bond.

This exhibition at the Palace Museum brings us fulfilment on more than one account. It is the culmination of the special relationship between Cartier and its trusted partners in Beijing. It is the recognition by one of the planet's most prestigious museums that Cartier creations constitute a heritage of immense historical value.

It is also a great personal pleasure, as I have a particular fascination for China's civilisation and its people. I come here regularly, perhaps four times a year, always in search of new opportunities and encounters, travelling to towns in the country's furthest reaches and even attempting to unravel the mysteries of the Chinese language.

Just as these mutual ties and influences are not new, so they grow stronger.

As early as 1888, there are records in Cartier's archives of lacquerwork objects, made for a prestigious patron, Vicomtesse de Bonnemain. Parisian antique dealers, including C.T. Loo on Rue de Courcelles, supplied Cartier with laques burgautés. Chinese images of animals, colours and symbolic icons have been introduced by Cartier to its jewellery designs. In the 1920s, powder compacts, vanity cases and even small clocks were incrusted with this shimmering inlay, with its evocations of faraway lands. Coral and jade made daring colour combinations. Mythical creatures – chimerae, dragons, Fo dogs and phoenixes – came to life on jewellery and mystery clocks. Louis Cartier himself was an informed collector of these works. Other motifs, some symbolic – stylised ideograms, Yin and Yang – and others religious – pagoda roofs and divinities – transformed objects into precious talismans. Chinese aesthetics have held a fascination for Cartier that continues today. The dragon, alongside the panther, remains one of Cartier's most revered animals.

Jeweller of kings, king of jewellers, a familiar presence in the antechambers and ballrooms of palaces, all that remained was for Cartier to step through the gates of the Palace Museum and enter its vast halls, where the murmur of imperial treasures can still be heard. The room which the directors of the Palace Museum have chosen for this exhibition stands in the axis of the South Gate, affording a spectacular view of the entrance to the City where thousands of visitors throng each day. Fully aware of the privilege which this extraordinary showcase, one that will recount the history of Cartier, represents, I particularly wish to thank Mr Li Ji, Executive Deputy Director of the Palace Museum, Ms Ma Haixuan, Deputy Director of the Foreign Department, Mr Ma Jige, Deputy Director of the Exhibition Department, and Mr. Song Hai Yang, Researcher and Curator of the exhibition, for their dedication to this great project.

There can be no doubt that this exhibition will remain a milestone in the cultural ties between West and East, from Paris till Beijing, and between Cartier and its Chinese friends.

President and CEO, Cartier International

Bernard Fornas

东方文化的浸润
——卡地亚珍宝中的中国元素

    卡地亚是享誉全球的著名珠宝腕表品牌，一百六十多年来，卡地亚始终保持着卓越的创新本质和超凡的精致工艺，珠宝大师们遵照严格标准，挑选最上乘的钻石、珍珠、翡翠、玉石、玛瑙、祖母绿、红蓝宝石等，经过精心设计和雕琢，赋予这些珍宝以灵性。卡地亚作品所拥有的文化内涵和典雅、高贵气质闪耀着恒久魅力，被奉为经典，成为一百多年的时尚。在卡地亚经典藏品中，有相当一部分作品受到中国艺术影响，这种影响并非孤立的，而是有着深厚的历史渊源，主要得益于创始人以及几代设计师对中国文化的浓厚兴趣，以及对中国元素的吸纳与融合。

一 20 世纪前欧洲大陆的中国风格

    20 世纪前欧洲大陆流行的中国风格，实际是中国传统文化所表现出的思想特点和艺术特色在西方设计领域中的表现。关于中国风格，《不列颠百科全书》的解释为："指 17 ～ 18 世纪流行于室内、家具、陶瓷、纺织品和园林设计领域的一种西方风格，是欧洲对中国风格的想象性诠释。"日本《新潮世界美术辞典》对中国风格的定义是："中国风格指西方人在远东特别是中国文物上寻求启迪和表现源泉的艺术倾向，以及由此产生的作品。又称'中国样式'和'中国趣味'。"[1]

1 盛世传播和大航海时代

    所谓一种文化对另一种文化产生影响，一般来说需满足两个基本要素。第一，社会经济发达，文化相对强势；第二，具备传播渠道或媒介。中国与欧洲贸易的加强与发展为文化传播创造了条件。历史上，每一次中国风格的传播，无不依托于社会经济的繁盛。西汉张骞出使西域以来，"大汉"、"盛唐"凭借丝绸之路掀开了阻隔东西方交往的珠帘。中国使者、商人接踵西行，西域使者、商人也纷纷东来。1275 年，马可·波罗随他的父亲和叔父来到中国，不仅觐见了元世祖忽必烈，而且在中国生活了 17 年（1275 ～ 1292 年），足迹遍及华夏。1292 年夏，马可·波罗随父亲、叔父奉命护送蒙古公主阔阔真远嫁波斯，三年后回到欧洲。他在参加与热那亚的战争中被俘，在狱中口述完成了《马可·波罗游记》，详细描绘了"天朝上国"的富饶物产。书中关于中国富庶繁荣，城市中丝绸、胡椒、金银、珠宝丰富，商路畅通，驿站完善，文化昌明的叙述对欧洲产生了巨大影响。

15世纪，人类历史迎来了波澜壮阔的大航海时代，最初的30年，在亚欧大陆两端的东西方，几乎同时向海洋进军。东方以中国郑和下西洋为标志，西方以葡萄牙王子亨利沿非洲西岸探索为代表。郑和下西洋的15世纪初，中国正处于明朝初期的"永乐盛世"，这是中国古代历史上最辉煌的时期之一。明成祖朱棣在太祖朱元璋恢复经济和生产的基础上，使经济得到进一步发展。其中松江成为全国的纺织中心，景德镇成为全国制瓷业中心，永乐时期也被誉为中国"瓷器的黄金时代"。郑和下西洋除了外交和军事使命外，发展海外贸易、传播中华文明也是重要使命之一。在郑和下西洋这一壮举的带动下，通过朝贡贸易、官方贸易、民间贸易等形式，丝织品、陶瓷、漆器等艺术品随着货物一起输出海外。当葡萄牙航海家达·伽马沿非洲西海岸绕过好望角，抵达东非海岸时，当地人就告诉他几十年前中国人曾几次来到这里。葡萄牙人在阿拉伯领航员的帮助下，沿着郑和船队开辟的航线顺利到达了印度。1517年，葡萄牙人托梅·皮雷斯受国王曼努埃尔一世派遣，作为欧洲使臣从广州登陆踏上中国的土地。此后，欧洲商人、传教士、外交使节纷至沓来，当他们回到欧洲时，带回的不只是商品、艺术品，其中的许多人还仿学马可·波罗撰写游记或回忆录介绍中国，如：葡萄牙人达·克路士的《中国志》、西班牙修道士马丁·德·拉达的《记大明的中国事情》（又名《中国札记》）等 [2]，这些书籍对于欧洲人了解中国起了重要作用，并且引起了欧洲人对"中国风格"的诠释。

17～18世纪中国风格装饰艺术在欧洲盛行。1600年，英国为了拓展东方贸易，创建了东印度公司。1613年，英国在印度西部的苏特拉设立贸易站，不久，又在印度东南部的马德拉斯建立商馆。1698年，东印度公司买下了位于孟加拉湾恒河口岸的加尔各答，设立贸易总部，除了把印度的粮食和工业原料源源不断地运回英国外，中国商品也吸引了东印度公司。东印度公司在广州进行季节性贸易采购，将大量中国商品、艺术品运往欧洲，从中获得丰厚利润。此后，荷兰、丹麦、法国、瑞典等国相继成立东印度公司来华或通过印度与华贸易，除采购香料、茶叶、生丝、药材等原料以外，还购买陶瓷器、丝绸、漆器、紫砂、绘画、壁纸、雕刻品、家具、屏风等。这些器物或艺术品的造型和纹饰对于欧洲人而言充满浓郁的异国情调，有别于欧洲教堂艺术、古典艺术风格，激发了皇家、贵族、学者等收藏者的欲望，"中国热"在欧洲渐渐升温，导致中国外销艺术品供不应求，终于迎来了欧洲"中国风格"设计的黄金时代。

2 康熙盛世与太阳王路易十四

康熙皇帝和法王路易十四分别是东、西方历史上功绩卓越的统治者，他们为推动东西方文化交流做出了巨大贡献。1667 年（康熙六年）7 月康熙帝亲政，凭着睿智和胆略，开创了盛世王朝。通过他的勤政、治国，奠定了中国辽阔版图，使中国成为当时世界上领土面积最大、最强盛的帝国。康熙时期民族多元和谐统一，从康熙二十年到康熙六十一年（1681～1722 年），41 年间中原地区社会稳定，经济发展，文化繁荣。康熙天性好学，终身手不释卷。他不但继承、学习汉文化，而且引进西学，精通数学、物理学、医学、化学、光学、测绘学、地图学。他注重与法国、意大利等西方国家的联系，与西方传教士频繁交往，使他们人尽其才，效力皇朝，对推动中西文化交流做出积极贡献。

路易十四施政时期恰逢中国"康熙盛世"，是中国文化、艺术的繁盛时期，陶瓷、书画、家具、各类工艺品、装饰艺术等异彩纷呈，各领风骚。1697 年，法国来华传教士白晋的著作《康熙帝传》在巴黎刊行，掀起一阵中国热。康熙朝民富国强，政治、经济、文化、艺术的发展对法国具有强烈的吸引力，王室、贵族首先表现出对中国艺术品的极大热情。路易十四的首相、枢机主教马萨林是最具代表性的人物之一，他收藏的扇子、丝织品、橱柜等中国艺术品非常丰富 [3]。他对中国艺术品的爱好，带动了法国贵族、皇室成员，也影响了法王路易十四。路易十四酷爱中国艺术品，辟凡尔赛宫镜厅专藏瓷器。1670 年，路易十四为他的情妇蒙特斯潘夫人修建特里亚农宫，该宫仿照中国建筑风格，外表用法国仿制中国瓷生产出的软瓷装饰，因此又称为"瓷宫"，成为当时颇具特色的时尚建筑。路易十四时期，凡尔赛宫经常举行东方艺术情调的化妆舞会，皇家乐队用笙、笛、锣等中国乐器演奏音乐，国王、王后、贵妇们身着中国丝绸、刺绣服装出席舞会成为时髦的事情。1700年 1 月，凡尔赛宫举行盛大舞会，路易十四别出心裁的乘坐中国式的轿子出场，足见他对中国式器物的喜爱。在路易十四时期的"宫廷动产总清单"中，记有许多中国造物品 [4]。在凡尔赛宫、康边与圣日耳曼昂莱宫王后的房间，有整套的中国家具，枫丹白露王太子办公室摆放着中国的扶手椅和折叠椅，甚至连国王的仆人家中，也有中国的漆桌、花瓶和挂毯。

随着欧洲各国与中国的直接贸易进一步增多，大量中国艺术品登陆欧洲，受到上流社会权贵、富绅的喜爱，在家中摆放中国的瓷、漆工艺品，或悬挂中国水墨画成为欧洲社会时尚，18 世纪达到高潮。

3 中国艺术品的影响

大部分没有到过中国的欧洲人是通过外销艺术品认识中国、感知中国文化的。17 世纪和 18 世纪法国、德国、西班牙、葡萄牙、荷兰以及英国等欧洲国家，在社会生活的方方面面体现了艺术家们从中国艺术品中所获得的创作灵感。中国艺术品所传达的儒雅、温和、自由、崇尚自然的理念，影响遍及绘画、雕塑、服饰、家具和室内装饰、建筑、园林等领域。中国龙、凤、麒麟、虎、鹿、蝴蝶、蝙蝠、梅兰竹菊、荷花池塘、岁寒三友、牡丹、芭蕉等动植物纹饰，还有山水园林、风俗故事、仕女婴戏、刀马人物，以及云纹、水纹等连续图案被作为装饰元素提取出来，用于装饰设计。正是由于中国艺术品师法自然、富于联想的自由形式和抒情达意，打破透视、比例及对称等原则，带给了欧洲全新的设计理念，激发了众多欧洲艺术家的想象力，在收藏的带动下，大约200年的时间里，模仿中国风格的装饰形式在欧洲流行。

织绣 中国丝织品具有的柔软飘逸、富有弹性、闪烁美妙光泽和变化多样的花纹图案特质，使其最早成为中外文化交流的载体。输入欧洲的中国丝织品以面料坯绸、刺绣和手绘丝绸为主，从最开始欧洲贵族的奢侈品，逐渐"平民化"，从制作服装，到丝质地毯、挂毯，甚至代替壁纸被用于装饰墙面，一直备受青睐。欧洲宫廷服饰，受中国刺绣影响，普遍装饰刺绣花纹、折裥、蝴蝶结等，甚至连贵妇们的高跟鞋也用中国丝绸、织锦、刺绣作鞋面装饰。法王亨利四世时期，宫廷内东方丝绸、刺绣盛行一时，宫廷刺绣匠师瓦尔特和园艺家罗宾创建了刺绣协会，向宫廷刺绣匠师们提供具有东方风格的刺绣图案和设计风格，并仿制中国风格的丝绣[5]。17世纪下半叶，中国手绘丝织品成为法国最时髦的样式。俄罗斯是欧洲与中国接壤的国家，从彼得大帝时期开始，中国风格就成为俄罗斯艺术中的突出现象之一。在圣彼得堡很多宫殿里可以看到中国陈设品或是具有中国样式的仿品。缅什科夫宫内还有罕见的中国丝绸替代壁纸包墙，丝绸上所画题材取自于中国小说《封神演义》[6]。

陶瓷 15世纪，中国陶瓷已见于欧洲文献。但16世纪前，输入欧洲的中国瓷器极其有限，而且价格十分昂贵。17世纪中期，中国陶瓷在欧洲仍被视为珍玩，只有欧洲宫廷有较多陈设和收藏，作为贵族装饰房间的艺术品。如：波兰王约翰三世在宫廷内设置专门陈列瓷器的中国厅；英国女王玛丽二世在荷兰居住时就喜欢用瓷器装饰别墅，到英国后在汉普顿宫仍以陈设瓷器为装饰[7]。随着清政府开放通商口岸，允许外国在广州开设贸易机构，进口瓷器大量增加，使得欧洲瓷器在逐渐摆脱单纯的装饰之用，餐具、咖啡具等日用瓷器需求日增，最终发展到通过欧洲商人向中国订制瓷器。明万历时期，中国外销瓷生产和销售已达到相当大规模。17世纪晚期，中国外销瓷的欧洲市场已经成熟，为了适销对路，江西景德镇、福建、广州等地窑厂接受欧洲商人的订单，按照欧洲人对造型、尺寸、纹饰的要求或图纸，依样生产，再销往欧洲。清末刘子芬所撰《竹园陶说》中记载："海通之初，西商之来中国者，先至澳门，后则径趋广州。清代中叶，海舶云集，商务繁盛，欧土重华瓷，我国商人投其所好，乃于景德镇烧造白瓷，运至粤垣，另雇工匠仿照西洋画法，加以彩绘，于珠江南岸之河南，开炉烘染，制成彩瓷，然后售之西商。"这其中最著名的是各种纹章瓷，也称徽章瓷。欧洲一些政府部门、皇家贵族、军队在中国订制瓷器，绘上代表他们的徽章。现存里斯本的葡萄牙曼努埃尔一世（1469～1521年）的青花纹章瓷执壶被认为是欧洲最早向中国订制的纹章瓷器[8]。

随着瓷器贸易的扩大，欧洲对中国瓷器需求增加，刺激了欧洲瓷器生产。早在16世纪末，意大利佛罗伦萨陶工在佛朗西斯一世大公的支持下就仿照中国青花瓷，烧制釉陶产品，又称梅迪西瓷，至今欧洲博物馆还有少量收藏。这种梅迪西瓷器的色彩和纹饰明显仿学中国的青花瓷器，白地蓝花，多以花果纹装饰，有别于16世纪欧洲彩陶装饰多为《圣经》中的场景或神话故事题材。由于是低温陶瓷，所以无论胎质还是釉质与中国青花瓷器相距甚远。17世纪，法国里昂、德国迈森、荷兰代尔夫特、英国的斯塔福德纷纷仿制中国瓷器或宜兴紫砂陶器。荷兰代尔夫特窑烧制的锡釉陶器，以仿制中国青花和五彩瓷而闻名天下。代尔夫特最初的"希诺兹利"（中国风格）样式，十分接近中国瓷器纹样，它以白地蓝彩纹样为主，近似中国青花瓷，装饰纹样多是龙、凤、狮、亭台楼阁、仙人、庭院花枝、山水风景等中国青花纹样，当时欧洲人称之为"支那的形象"，认为代尔夫特陶器最近东方风格。代尔夫特陶器虽然充满东方情调，装饰风格摹仿中国瓷器，但造型完全按欧洲人的生活习惯制作，形成欧洲大陆所特有的奇异风格。18世纪，德国迈森生产的瓷塑"中国皇帝和随从"，虽然形象和服饰都很欧式，体现出欧洲人形式概念和中国人含义的混合，在当时的欧洲轰动一时，成为最受欢迎的产品 [9]。

漆家具和壁纸　16世纪末，已有少量中国填漆或漆绘家具出现在英国、法国等一些富商住宅中。路易十四以后，随着中国风格的流行，英国、法国、德国、荷兰等国兴起仿制中国漆器之风。各种橱柜、桌椅、箱盒、屏风等，采用黑漆描金图案，装饰华丽，广受欢迎。法国马丁兄弟四人，德国的司徒华骚均以仿制中国漆器著名。而英国著名家具设计师托马斯·契潘达尔于1754年出版了一本关于家具设计的著作《绅士和细木工导师》，收录了帕拉迪奥建筑式样、哥特式和中国家具图案。这本书对中国风格设计在欧洲流行发挥了重要作用。大约在1755年，托马斯·契潘达尔制作了中国式床（宽2.41米，收藏于伦敦维多利亚和阿尔伯特美术馆）。该床采用中国式的四柱形式，结合英国特有的尖形顶盖，风格看起来有些怪异，但确是英国工艺"中国风格洛可可化"的代表作 [10]。一件意大利1740年的中国式抽屉柜，上面的图案是中国传统的山水画，可以看出当时欧洲家具受到了中国艺术的巨大影响。事实上，漆家具上的描金图案、华丽纹样对欧洲设计的影响，不仅限于家具、陶瓷、金属工艺、纺织品、壁纸、壁毯以及室内设计的图案，均借鉴中国漆家具纹饰。

欧洲人热衷于用壁纸装饰房间，16～17世纪的海上强国荷兰、英国开始输入中国壁纸，随后有人开始仿制。18世纪带有人物、禽鸟、花卉图案的中国壁纸与中式家具的室内装饰珠联璧合，受到欧洲普遍欢迎，因而刺激了欧洲壁纸生产。设计师从中国艺术品上择取花枝、蝶鸟、庭院、山水、四季清供等装饰纹样作为设计元素借用在壁纸（布）上，疏朗、优雅的东方情调广受欢迎。1754年，英国人杰克逊在巴特西建厂生产壁纸，并选择中国山水画为装饰纹样 [11]。随着仿制中国风格的壁纸越来越多，出现了中英合璧、中法合璧的风格，壁纸图案或追求精美华丽，或疏朗淡雅，赋予中国主题装饰新的想象和诠释。

绘画　中国风格对欧洲绘画艺术的影响，也是从欧洲人对中国艺术品的兴趣开始的。画家将中国艺术品纳入绘画创作当中，如荷兰画家奥西亚斯·贝特尔1608年创作的《装着樱桃和草莓的中国瓷碗》和皮特·布尔 1670 年创作的《有地球仪和豪华器皿的静物》，都将中国瓷盘作为创作对象进行描画。18 世纪洛可可绘画的代表人物华托与布歇都对中国风格充满兴趣，华托曾经多次画过中国人物形象。而布歇不仅设计了《中国皇帝的召见》、《中国婚礼》、《中国风俗》、《中国花园》等题材的九幅壁毯图画，而且他的素描《吹笙女子》完全是中国仕女形象。法国贝桑松博物馆、巴黎马蒙达博物馆、荷兰鹿特丹伯门斯－凡·毕尼根博物馆至今还收藏有布歇的《中国皇帝上朝》、《中国花园》、《有中国人物的风景》、《中国渔情》等油画作品 [12]。布歇是一位在素描、色彩方面有很深造诣的画家，这些画中的人物造型完美，既显示了古典绘画的影响，又体现了洛可可风格的装饰趣味和新颖的东方情调，表现出一种追求风雅、愉悦、舒适、逍遥的生活态度。

4 中国建筑装饰、园林艺术的影响和洛可可风格

中国建筑和园林艺术在欧洲被接受，以至于流行，其原因是多方面的。以利马窦为代表的一些到过中国的传教士盛赞中国庭园设计，认为中国不囿于形的庭园形式，较欧洲几何式庭园更美，更自然恬静。值得一提的是由荷兰阿姆斯特丹出版的两部带有插图的著作对引发欧洲人对中国园林、装饰风格的兴趣起了不容忽视的作用。一部是 1655 年出版的荷兰人尼霍夫的《中国旅行记》，配有150 幅插图；另一部是 1667 年出版的基歇尔的《中华文物图志》，书中真实形象地记录了中国，带给欧洲人对中国装饰的直观认识。1749 年，法国天主教耶稣会传教士、画家王致诚创作的《中国皇帝游宫写照》（1743 年）发表在《耶稣会士书简集》，将圆明园介绍给欧洲，引起法国人对圆明园的心驰神往。后来王致诚又将中国画家唐岱、沈源所绘圆明园 40 景图副本寄往巴黎（现藏巴黎国家图书馆），成为法国人模仿中国园林建筑的范本 [13]。同时，欧洲人从中国瓷器、漆器，以及丝绸上的亭台、庭院等建筑园林景观装饰纹样中获得启发，仿学中国建筑、园林形成风气。

欧洲人以新奇浪漫的心态接受中国建筑、园林艺术，在一些英国花园里出现了中国式凉亭。最著名的当属威廉·钱伯斯在 1757～1762 年之间为威尔士王妃奥古斯汀公主改造位于伦敦西南部的邱园，在园中按照中国式样设计湖石假山、瀑布岩洞、曲径花草等景观，并且设计建造了一座中国式的九层宝塔，成为荷兰、德国等欧洲国家仿效的范本。1781 年，德国卡塞尔伯爵在其领地威森斯泰恩修建"中国村"，并命名为"木兰村"，几乎所有建筑都是中国式平房，其中的宝塔和魔桥至今尚存。仿效借用中国建筑中的翘檐尖顶、楼台亭阁、水榭宝塔成为 18 世纪的时尚风气，德国波茨坦附近三苏栖花园中的中国凉亭和中国趣味的小建筑，以及瑞典王后岛上的"中国宫"都留下中国园林艺术影响的痕迹 [14]。

中国园林师法自然的意识、闲适恬静的情趣影响了欧洲艺术流派，表现为"洛可可"的自然情怀及多元的艺术风格。洛可可式的工艺装饰代表一种纤巧、精美、华丽的艺术风格和样式，被认为

是意大利巴洛克艺术传入法国的遗留物，还有一种观点认为和中国园林艺术中的湖石假山有关。不管怎么说，中国园林以及中国师法自然的艺术理念，给予欧洲洛可可式的建筑装饰和室内设计风格相当丰富的滋养[15]。那些描绘自然景物的装饰图案，墙面圆弧设计和曲线的应用，成为洛可可风格的最大特色。日本学者小林太市郎认为应将"洛可可式"称为"中国·法国式"。实际上中国风格的影响不仅限于法国，而是伴随着地理大发现和世界市场的建立出现在整个欧洲的文化艺术现象，直到19世纪末新艺术运动兴起，都不能摆脱中国人"师法自然"的思维，只不过不是直接模仿自然形式，而是透过选取与剪裁重新完成一种构思。19世纪末是卡地亚成长、发展时期，中国风格的遗韵，新艺术运动的兴起无疑对卡地亚设计大师们追求时尚的设计产生了重要影响。

二 卡地亚珠宝中的中国元素

卡地亚的传奇始于1847年，从28岁的路易·法朗索瓦·卡地亚接手师傅阿道夫·皮卡尔位于巴黎蒙特吉尔大街29号的珠宝首饰店开始，至今已逾160年。1846年，当路易·法朗索瓦·卡地亚还是一名珠宝工匠时，就将其姓名缩写字母"LC"设计成环绕一颗心的菱形图案注册了自己的商标[16]，由此显示出他超凡的商业意识。卡地亚家族有博览群书的传统，不仅收藏艺术史丛书，而且游历世界，收藏埃及、中国、日本、印度、俄罗斯以及阿拉伯和波斯等国家和地区的文物、艺术品。我们在翻阅卡地亚发表的一些历史资料时，见到一张曾任卡地亚高级珠宝总监贞·杜桑女士拍摄于1925年的照片[17]。

贞·杜桑女士坐在中国彩漆屏风前，时尚优雅，充满魅力。照片中的彩漆屏风应该是中国清中期的精品，屏风上的华丽装饰和山水人物清晰可见。游历世界，收藏丰富，博学广识，使得卡地亚的艺术创作极具传统内涵，散发着高雅的文化气质。几代设计师始终追求耳目一新的感受，从来自异域情调的艺术中吸取养料，寻找灵感，融合创意之美，模仿并超越它们。卡地亚设计师们巧用金、银、各类宝石组合重新诠释美丽和高贵，将华贵和优雅推向至高境界。

从卡地亚20世纪二三十年代制作的一系列作品可以看出，这时期卡地亚的珠宝设计受19世纪以来欧洲流行的中国装饰风格影响，反映出设计师对中国艺术的迷恋，极力吸收中国元素，作品渗透出浓浓的中国韵味，令世界着迷。这些带有中国文化韵味的作品，受玉雕、漆雕和丝织品装饰纹样的影响最大，具有鲜明的特点，主要表现在以下几个方面：

1 直接植入中国艺术品

卡地亚珠宝设计师大胆借用中国艺术成品，直接植入新作品的设计之中。根据所用成品的材质、形状，佐以欧洲趣味的珠宝装饰，充满浓厚的东方韵味，成为卡地亚颇具特色的珠宝系列之一。1912 年巴黎卡地亚借用一件中国清代女子梳妆用的花卉纹翠玉粉盒，加上以橄榄形、凸圆形红宝石和玫瑰式切割钻石制作的盒扣，改制成糖果盒 (BS 05 A12)，是一件卡地亚早期中国风格的作品。

1921 年，巴黎卡地亚的作品翡翠吊坠 (PE 15 A21)，取材于一件上等中国清代荷花纹翡翠佩，辅以符合欧洲人欣赏口味的红宝石和玫瑰式切割钻石，成为一件新的精致挂饰。娇艳的红色与翡翠绿搭配成标准的中国色。1924 年，巴黎卡地亚的设计师根据客户提供的一件中国清代翡翠龙形带钩，以蓝宝石、圆形或单面切割钻石、黑色珐琅为配饰点缀，设计制作成充满神秘色彩的龙形胸针 (CL 80 A24)。1935 年，巴黎卡地亚将一件中国文房用具白玉莲花笔洗，加上黑曜石底座，点缀雕花珊瑚，改制成高雅别致的烟灰缸 (SA 04 A35)。

上述四件巴黎卡地亚的作品都是直接利用成品，拿来为我所用。以原物为主要设计元素，设计师极力保持植入艺术品的原貌和文化气息，稍加修饰，成为欧洲人更易接受和欣赏的装饰品，创意新颖，构思巧妙。1928 年巴黎卡地亚制作的粉盒 (PB 20 A28)，上下两面为两块透雕翡翠，盒盖正中嵌蓝宝石金线扭丝纹，盖钮套青金石环，盒壁为蓝绿色连续图案掐丝珐琅，精巧华丽。1930 年，伦敦卡地亚将 21 块中国 19 世纪圆形翠玉，模拟清代光绪铜钱形制，并雕有"光绪通宝"字样，钱孔内嵌红宝石，以金环连接成腰带 (JA 26 A30)。其设计思想显然受明代玉带的影响，具有典型中国色彩特征和独特的装饰效果。

卡地亚还有一些具有典型欧洲风格的珠宝饰品，中国艺术品在其中并不是作品的主体，只作为设计元素之一融入作品设计中，这些作品呈现西方艺术风格却夹裹着浓浓的中国韵味。

化妆盒 (VC 14 A26)，设计师将化妆盒设计成中国枕形，枕面黑色珐琅装饰"卍"字锦纹，两块五蝠捧寿漆牌嵌在枕形化妆盒的两端。带有吉祥寓意的五蝠捧寿漆牌，原为两块漆嵌件，显然，设计师是根据这两块漆牌构思设计了这个中国风格的化妆盒，嵌上玫瑰式切割钻石，更显富丽华贵。佛犬镇纸 (DI 08 C27)，镇纸为墨玉质地，镇纸顶部镶嵌的佛犬，实为一件中国清代晚期的圆雕象牙瑞兽。这种瑞兽在中国古代艺术品中常见，形象一般为圆眼、宽鼻、阔嘴、垂耳、狮爪，通常呈俯卧状，有平安祥瑞之意。在这件作品中，象牙与墨玉相配，辅以蓝宝石珠、红色珐琅云头纹装饰，既和谐又有几分神秘感。腰带 (JA 05 A28)，带扣为一块中国翡翠玉牌，配合绿松石、蓝宝石和蓝色丝带，和谐典雅，现在看也是一件极具时尚品位的经典作品。印章表胸针 (WB 34 A29)，由金、铂金、钻石、红宝石珠、玛瑙、黑白珐琅等材质制作，镶嵌了一枚中国 19 世纪翡翠狮钮印章，印纹部位嵌接一块有黑白珐琅制作的长方形表，构思非常独特。

除珠宝饰品之外，各种材质的高档钟表也是卡地亚特别擅长制作的，大胆植入中国艺术品使不少卡地亚高档钟表成为经典之作。带飞返指针鲤鱼时钟 (CS 11 A25)，以中国清代圆雕青白玉双鲤鱼扇形飞返式时钟，黑曜石底座，有黄金、铂金、珊瑚、珍珠、钻石等细部装饰，是一件中国风格鲜明，又具全新创意的作品。喀迈拉魅幻时钟 (CM 23 A26)，在以金、铂金、珍珠、钻石和黑色玛瑙、黄水晶（表盘）、珊瑚等材质制作的钟表上，植入中国传说中的神兽，除富贵华丽、工艺精湛之外，更具一种神秘的力量。喀迈拉为英文 chimera 的译音，传说中的神兽、怪物。此钟的设计灵感来自于法国路易十五、路易十六时期的时钟设计，通常将钟表的机芯置于动物的背部。该钟是卡地亚最为名贵的标志性藏品。神像报时魅幻钟 (CM 04 A31)，以一尊 19 世纪中国白玉观音立像为主体，两边为同时期白玉瑞兽和青玉笔筒，笔筒内插一支珊瑚，分别镶嵌绿松石和珍珠。观音立像前装置八角形无色水晶表盘的报时钟，绿松石镶边。观音像玉质洁白温润，工艺考究，本身就是一件上乘的玉雕作品，经卡地亚设计师的再创作，视觉效果错落有致，色彩鲜艳又不失和谐，普度众生的观音和象征平安吉祥的瑞兽，更反映出深层的中国文化内涵。瑞兽时钟 (CS 05 A43)，将 1928 年拆解的一座时钟上的 19 世纪中国红珊瑚龟驮瓶雕件重新使用，配以白色玛瑙底座和钟盘，钻石时针和时刻，红珊瑚龟驮瓶雕件镶嵌钻石装饰，整个作品色彩温润，艳雅脱俗。龟、瓶组合在中国吉祥文化中寓意平安长寿。

卡地亚的中国风格作品中，有一些利用饰件或器物局部设计加工的作品，同样不失浓郁的中式风格韵味。如插屏式座钟 (CDB 23 A26)，钟盘显然是利用一件中国白玉浮雕插屏的局部抑或是残品制作的，从雕工工艺看，应该是清代中期作品。局部的玉插屏被卡地亚的设计师包上镶嵌着红珊瑚的黑色珐琅边框，欧式表针，以及金、铂金、钻石、螺钿、玛瑙等细部处理，成为呈现中国意韵无可挑剔的完整作品，保持了卡地亚始终如一的富贵典雅风格。棕榈叶棘爪式扣针 (CL 244 A25)，利用中国清代玉佩的残片制作，玉片上还留有"福在眼前"的中文吉祥语 [18]。经过卡地亚设计师的巧思妙想，以红宝石、钻石、金、黑色珐琅的修饰搭配，制作成既有印度风格，又有中国意韵的精美作品。

2 择取中国文化元素

吸收世界各国艺术精髓融入卡地亚珠宝、饰品设计，是卡地亚珍宝的又一个特色。卡地亚珠宝中相当一部分作品从中国艺术品绘画、陶瓷、漆器、丝绸、玉器等中国文化载体中择取设计元素，并经过二次创作扩展了这些文化元素的意义。中国文化的儒雅气质与卡地亚珠宝的华丽富贵完美结合，使得卡地亚珠宝充满古老文化的内涵，而散发着时尚的气息。

阴阳吊坠 (NE 19 A19)，为巴黎卡地亚 1919 年制作。将中国的太极两仪图形引用在首饰设计中，以黑色玛瑙和钻石制作一对太极两仪图纹的球形挂坠，镶嵌红宝石珠，并以红宝石、翡翠装饰，赋予了中国"太极"是派生万物本源的哲学思想新的表达形式。"寿"字耳坠 (EG 24 A26) 和弥勒佛耳坠 (EG 11 A28)，分别将变化了的中国"寿"字和佛教中的弥勒佛形象作为装饰元素运用在首饰制作中，配合西式的装点，设计颇具创意。"寿"字在中国作为设计元素不仅历史悠久，而且变化多样，常用于象征长寿和吉祥寓意的作品中。民间有弥勒佛的布袋能将灾难带走的说法。

卡地亚早期的许多作品将中国不同时期的龙纹、龙形融入在设计中，形成"喀迈拉"系列。1923 年巴黎卡地亚制作的喀迈拉棘爪式扣针 (CL 260 A23)，仿游龙戏珠的形式。而根据龙形设计的喀迈拉手镯更成为卡地亚珠宝中的经典之作。喀迈拉手镯 (BT 109 A28)，为二龙戏珠形式，龙口衔珠相对，珊瑚雕制的龙头与瓜棱式绿色琉璃珠搭配成典型的中国颜色，金、铂金、钻石、蓝宝石、蓝绿黑色珐琅等多种材质的组合，工艺极为精致考究，成为卡地亚手镯中具有鲜明东方风格的经典作品。铂金喀迈拉手镯 (BT 064 A29) 是卡地亚喀迈拉系列第一款全铂金宝石手镯，以铂金、梨形或圆形单面切割及法式切割钻石、凸圆形和面弧形蓝宝石、无色水晶制作。龙头相对，形神与中国龙无二。1922 年巴黎卡地亚的两件作品将龙纹和螭虎纹分别运用在挂饰和胸针上。胸针 (CL 258 A22)，黑色珐琅环嵌钻石制作的螭虎纹，显然受中国玉器或青铜器中的螭虎纹影响。卡地亚另一件龙纹化妆盒 (VC 69 A27) 上采用的龙纹应该是清代晚期的龙纹形式，这种龙形图案一般在中国的纺织品中较为多见。

吊坠 (NE 03 A22)，设计灵感来源于中国商周时期的挂饰，主体部分圆柱形红珊瑚上镶嵌钻石制作的行龙纹，非常生动传神。形式设计简洁，制作精致。中国西周时期的琢玉技术较前代有了明显提高，水晶、玛瑙、绿松石、玉石等不同硬度的美石可以随意搭配使用，常见的以璜、佩、管、珠等串连，水晶、玛瑙、绿松石与玉石搭配组合，形制丰富多变，颜色绚丽缤纷，彰显华贵之气和强烈的装饰效果。

卡地亚还有一些作品模仿中国的花篮、花瓶图案。分别由纽约、巴黎卡地亚制作的两个胸针 (CL 47 A27, CL 48 A28) 和两个化妆盒 (VC 68 A27, VC 71 A28)，均选用中国瓷器上的花篮和花瓶题材设计，具有浓郁的中国风格。

化妆盒 (VC 67 A24)，两面为木质板面，多材质雕嵌中国人物故事图。两挎刀英雄相对施礼，绵羊环绕，祥云飞鹤，一派祥和气氛。上下接绿松石镶嵌钻石团"寿"字装饰，盒顶部钻石制龙戏珠，都是典型、饱满的中国元素。主要画面使用珊瑚、绿松石、螺钿、青金石、孔雀石、玛瑙等，制作工艺与中国的百宝嵌类似。墨水瓶 (DI 07 C27)，是 1927 年纽约卡地亚的作品，形制仿中国的香炉，瓶盖雕缠枝花纹、木质底座、底座与瓶体的连接件等都是典型的中国元素。还有一件颇具代表性择取中国元素的作品，即 1928 年巴黎卡地亚制作的化妆盒 (VC 72 A28)，它的主体画面设计是根据路易·卡地亚收藏的一件中国康熙时期的五彩瓷盘纹饰 [19]。这是一幅标准的中国仕女图，画面为庭院一角，画中女子执折扇两手相交抱于胸前倚坐在石桌旁，优雅柔美。身后的苍松翠竹，与石桌上的瓶梅，正好和为"岁寒三友"，松、竹经冬不凋，梅花耐寒开花，寓意坚强的品质。

3 带有典型中国特色的用材和工艺

卡迪亚珠宝的选材用料既丰富又考究，可以看出设计师在设计作品时，非常注重处理材料与材料、材料与造型、材料与色彩之间的关系，各种材质协调合理搭配，金、铂金、钻石与各色宝石之间互为装饰的作用，微妙和谐，足见设计师的美学造诣和活跃思维。在这些择取东西方文化精华，完美表达西方人审美取向和极具东方魅力的作品中，翡翠、玉、珊瑚、玛瑙、水晶、绿松石等是常见的宝石用材，这些宝石也是中国传统艺术品中最常使用的材料。其中，中国是最早将玉、水晶、玛瑙用于装饰品制作的国家之一，可以追溯到新石器时期。玉石在中国被誉为美石，以纯净温润为上品，无论在早期的礼仪、等级、丧葬等制度中，还是后来的世俗生活中都发挥了重要作用，玉器不但具有直观美，而且被赋予了太多的人文和社会道德属性，以至于被"德"化、"神"化、"权"化，最终形成玉文化，历数千年而不衰。

水晶、玛瑙、绿松石是广义的玉石家族成员，它们在中国装饰艺术中具有不容忽视的重要地位。水晶，矿物学又称"晶石"或"水晶石"，是一种无色透明或半透明的大型石英结晶体矿物。色美性脆，呈现玻璃光泽。结晶完美的水晶晶体常呈六棱柱状，多条六棱柱状的结晶连接在一起，被称为晶族，非常美丽。在形成过程中混入铁、钛、锰等不同矿物元素，就会呈现不同的颜色，清明晶莹。

玛瑙，是一种胶状体矿物，一般为半透明或不透明质地，原石肌理可呈现多种颜色或条纹，具有坚硬、致密、细腻，形状各异，光洁度高，颜色美观且色彩丰富等特点，非常适合作为工艺品进行雕琢加工。绿松石是铜和铝的磷酸盐矿物集合体，色彩娇艳柔媚，质地细腻、柔和，硬度适中。因所含元素的不同，颜色也有差异，以不透明的蔚蓝色最具特色。珊瑚是宝石中较为特殊的品种，为生于海底岩礁的有机宝石，中国的藏族更是对珊瑚喜爱有加，常常作为护身符和祈祷上天保佑的寄托物。我国新石器时期、商周时期出土的串饰，许多是以水晶、玛瑙、珊瑚制作的管、珠、环、璜、佩等组合连缀。翡翠，是一种以硬玉矿物为主的辉石类矿物集合体，透明晶莹、光泽喜人、硬而不脆，是公认的"玉中之王"。作为装饰品用材，中国清代的用量最大。由于皇帝、王公贵族的喜爱，尤其是受到清朝乾隆皇帝的推崇和慈禧太后的偏爱，翡翠身价百倍，成为玉中极品，有"皇家玉"之称。西方的工艺美术以陶、木、玻璃、金属工艺最为擅长，而卡地亚珠宝所展现的金、铂金、钻石与各色宝石的自如组合，说明卡地亚的设计师们一方面熟稔贵金属、钻石的特点，另一方面对玉、翡翠、珊瑚、玛瑙、水晶、绿松石的属性了如指掌，以无可挑剔的精湛工艺，使这些宝石在近现代珠宝艺术中焕发异彩。以玛瑙为例，时钟（CDB 07 A27），1927年巴黎卡地亚制作。钟体为玛瑙质地，透雕花鸟缠枝纹，仿中国清代器柄形制，钟面为黄水晶，钟盘与底座以青金石制成。黄色水晶、玛瑙与蓝色青金石为色彩学中的对比色，长方形八角钟面、方形底座与自然随意的玛瑙钟体形成对比，作品具有鲜明的中国风格，色彩天成，富于变化又和谐统一。

卡地亚具有中国风格的作品中，仿中国嵌螺钿工艺的作品颇具特色。嵌螺钿是髹漆工艺上的一种装饰手法，一般用杂色贝壳、螺蚌壳打磨成各种形状的薄片、细丝或大小不一的颗粒，拼粘成设计好的纹饰、图案，嵌在制好的漆胎上，再经层层髹漆，打磨，抛光，制成图案纹饰色彩斑斓的嵌螺钿漆器。中国的螺钿镶嵌工艺早在西周时期就已被用于漆器装饰了，唐代趋于成熟，明清时期随着漆器、家具等其他手工业的繁荣而达到高峰。嵌螺钿工艺具有装饰性强、纹饰色彩华丽炫目的特点，无论宫廷还是民间都广受欢迎。宫廷造办处有专门的工匠制作皇家御用嵌螺钿器，此外，北京、山西、苏州、广州等地都有生产，而扬州一带更是民间作坊林立、工匠云集[20]。嵌螺钿的工艺技法有硬螺钿镶嵌和软螺钿镶嵌之分，由此产生不同的艺术效果。品种极为丰富，大到门窗、屏风、家具，小到瓶、盒、杯、盘、碟、碗、筷子、羹匙、文房用具，都可以应用螺钿镶嵌工艺制成的山水人物、花鸟鱼虫、清供博古、几何图案等装饰。

卡地亚作品中一些以螺钿镶嵌图案的作品，与中国的嵌螺钿工艺极为相似，以红、黑色珐琅为底或胎，显然受中国漆器的影响。保持传统的中国式图案纹饰，但设计更加时尚，装饰更加华贵。作品表现为两种形式，一种是用螺钿作为装饰图案的底、背景或钟表盘，如凤凰粉盒及唇膏盒（VC 58 A25）和粉盒（PB 21 A27），盒面均以镶嵌螺钿拼粘的图案为底，上面再用金或红、黑珐琅等制成的龙、凤纹装饰。两个盒上嵌金丝折线纹和云头纹也都是择取的中国元素，是两件典型的中国风格作品。座钟（CBD 05 A26）和佛犬时钟（CBD 12 A26），都是以螺钿拼粘的图案作表盘，螺钿富有变化的天然色彩，华丽炫目。另一种是用螺钿拼粘纹饰、图案，如化妆盒（VC 54 A25），盒面的中心部分，黑色珐琅地镶嵌螺钿拼粘的祥云纹、蓝宝石镶嵌凤纹，金丝勾边。祥云的颜色变化自然微妙，有丹凤朝阳之意。座式烟盒（TB 07 A22），盒盖为螺钿拼粘的中国风景，四周衬托六角形或梅花形连续图案。座式烟盒（TB 11 A25），盒面设计成一幅展开的中国卷轴画，画面为螺钿拼粘的中国风景、人物故事，真是奇思妙想，独具匠心。三问座钟（CR 17 A30），钟体四面螺钿镶嵌人物和钟表时刻，色彩晶莹美奂，使厚重、规矩的长方体钟身取得富于变化的视觉效果，边框镶嵌的柱式翡翠和兽钮，更增加了座钟的中国韵味。

4 东西方文化交融的范例——鼻烟壶

卡迪亚品牌中国风格的作品中，有几件以中国鼻烟壶改制的作品，很有特点。鼻烟由西方传教士传入中国，最早出现在清康熙年间（1662～1722年）。在欧洲，鼻烟一般置于烟盒内，比较容易挥发、遗撒。为了适应中国人的着装习惯，方便储存、携带，各种材质、形制的鼻烟壶应运而生。最初宫廷造办处、御窑厂等御用生产机构设计制作，很快流行民间。鼻烟壶产生的年代恰逢中国清朝的康乾盛世，国富民丰，正是中国的工艺美术发展的辉煌时期，各工艺门类都发展到了相当高的艺术水平[21]。鼻烟壶有陶瓷、玻璃、玉、珐琅、水晶、奇石、竹木牙角等各类材质，浓缩各类艺术之精华，在很短的时间内由最初的实用品，迅速发展成为既方便实用，又可观赏把玩的综合性艺术品，成为中西文化交融的典范。例如利用中国鼻烟壶制作的充满东方文化内涵的香精瓶和一个打火机，分别为白玉、碧玉、黄玉、珊瑚材质的鼻烟壶，都是中国清中期鼻烟壶中的力作，经过卡迪亚设计师的精致修饰、加工，小巧玲珑，高贵奢华。碧玉香精瓶（FK 06 A25），瓶体为碧玉鼻烟壶，玉质润泽，光素无雕琢，可见深色自然纹理。盖和底座为蓝、黑色珐琅，包金刻花嵌红宝石装饰。翡翠香精瓶（FK 11 A25），瓶体为扁圆形翡翠鼻烟壶，质地细腻，色泽娇艳纯正，嵌蓝宝石珠和团"寿"字。瓶口包金。蓝、黑色珐琅瓶盖，金丝曲线装饰，盖顶嵌凸圆形蓝宝石。这两件香精瓶的金玉、蓝黑色珐琅搭配，使原本就素雅高贵的作品，增加了华丽的色彩。黄玉香精瓶（FK 05 A26），瓶体为花鸟纹黄玉鼻烟壶，形制平肩阔腹，玉质莹润，色泽均匀。配青金石嵌金丝瓶盖，扇形盖钮。黄玉与青金石的对比色搭配艳丽和谐。珊瑚香精瓶（FK 19 A26），瓶体为荷花纹珊瑚鼻烟壶。卡迪亚设计师基本保持了鼻烟壶的原状，只以少量钻石、金、珍珠、黑色珐琅点缀修饰，丝毫不失富丽华贵的气质。台式打火机（LR 22 A39），瓶体为白玉鼻烟壶，玉质温润洁白，通体刻文字装饰。配黑色珐琅包金瓶盖，嵌红珊瑚、钻石，明艳清雅，设计独具匠心。

卡地亚的设计始终紧紧抓住社会的流行趋势，这些具有浓厚中国韵味的作品，得益于 20 世纪前流行于欧洲的中国风格，特别是 18 ～ 19 世纪欧洲对中国文化的追崇。大航海时代为西方发现中国打开了大门，作为文化载体的各类艺术品则为卡地亚设计师提供了丰富素材和灵感来源，设计师们从中国文化中汲取营养，并以超凡的智慧、完美的工艺和无与伦比的想象力，诠释他们对中国文化的喜爱和理解。

三 结语

追求时尚在于理智而熟练的驾驭。从卡地亚这些充满中国风格韵味的作品可以感受到东方文化对西方装饰艺术的浸润和影响，无论是在作品中直接植入中国艺术品，还是设计中择取中国文化元素，才华横溢的设计师们始终表现出学习和吸纳不同民族艺术精粹的热情，将作品注入超越时光的美，连续一百六十余年奉献给世界。卡地亚的珠宝饰品设计各具特色而格调一致，简约时尚，经典浪漫。作品中蕴涵的人文情怀，拨动人心，总能激发人们的无穷欲望，被奉为凌驾于潮流之上的神话。

故宫博物院

宋海洋

注释：

1 2 袁宣萍：《十七至十八世纪欧洲的中国风设计》，文物出版社，2006 年

3 10 12 袁宝林主编，远小近、廖旸：《欧洲美术——从洛可可到浪漫主义》，中国人民大学出版社，2004 年

4 故宫博物院、凡尔赛宫博物馆编：《太阳王路易十四》，紫禁城出版社，2005 年

5 马良：《西方人眼中的东方丝绸艺术》，上海教育出版社，2004 年

6 韩显阳：《俄罗斯"北方之都"圣彼得堡的中国元素》，《光明日报》2007 年 9 月 5 日

7 9 11 14 张国刚等：《明清传教士与欧洲汉学》，中国社会科学出版社，2001 年

8 陈伟、周文姬：《西方人眼中的东方陶瓷艺术》，上海教育出版社，2004 年

13 周一良主编：《中外文化交流史》，河南人民出版社，1987 年

15 陈志华：《外国造园艺术》，河南科学技术出版社，2001 年

16 卡地亚珠宝公司编：《卡地亚》

17 上海博物馆编：《卡地亚艺术珍宝》，上海书画出版社，2004 年

18《香港佳士得》,1977 年 11 月 6 日销售目录

19 1979 年 11 月 25 日至 27 日摩纳哥苏富比拍卖行记录，第 101 号，艾伦·卡地亚提供的资料

20 沈从文著，李之檀编：《螺钿史话》，万卷出版公司，2005 年

21 故宫博物院编：《故宫博物院藏文物珍品大系——鼻烟壶》，上海科学技术出版社，2002 年

其他参考书：

《中外关系史译丛》第 4 辑，上海译文出版社，1988 年

［英］苏立文著，陈瑞林译：《东西方美术的交流》，江苏美术出版社，1998 年

许明龙：《欧洲 18 世纪"中国热"》，山西教育出版社，1999 年

陈伟、王捷：《东方美学对西方的影响》，学林出版社，1999 年

刘晓路：《世界美术中的中国与日本》，广西美术出版社，2001 年

严建强：《十八世纪中国文化在西欧的传播及其反应》，中央美术学院出版社，2002 年

李辉炳：《中国瓷器外销与外销瓷的生产》，《瑞典藏中国陶瓷》，紫禁城出版社，2005 年

张夫也编著：《外国工艺美术史》，高等教育出版社，2006 年

IMMERSION OF EASTERN CULTURE
—the Chinese Elements in Cartier Jewellery

Cartier is a world-renowned Maison, and has been synonymous with unparalleled craftsmanship and constant innovation for over sixteen decades. It's jewellery masters follow the strictest of standards in selecting the most precious gems, amongst them includes diamonds, pearls, jadeites, agates, emeralds, rubies and sapphires. They design and cut these treasures with great care and attention, bringing each piece to life. The brand possesses an everlasting charm of a rich culture and classic elegance, celebrated worldwide since its inception. A significant portion of the Cartier Collection has its roots in Chinese culture, which fascinated the Maison's founder and many subsequent designers after him. The Chinese elements have an indelible and enduring influence on Cartier, and its artists continue to draw inspiration from them.

1.The Chinese Style in the Continent of Europe before the 20th Century

Before the 20th century, the Chinese style in the European continent was a manifestation of the traditional Chinese ideology and artistic features in Western designs. It was defined by the Encyclopedia Britannica as 'one of the popular Western styles, which is a demonstration of the imaginative understanding of Chinese culture in the Western mind, in decoration, furniture, pottery, fabrics, and garden architecture between the 17th and 18th century.' The Japanese Fine Art Dictionary of the New World described it as 'the inclination of Western artists in drawing inspiration from Oriental, especially, Chinese antiques, or the artworks created in this way, also called the Chinese pattern or the Chinese taste.' [1]

1.1 Eastern Prosperity and the Western Age of Sail

The impact or influence of one culture upon another is generally based on two preconditions: 1.a stronger culture of more advanced economic and social development; 2. channels and media of cultural exchanges. In this instance, trade between China and Europe made way for the transmission of culture. Throughout history, all spreading of Chinese culture was the result of booming economy and thriving society. Since Zhang Qian's visit to the west, the Silk Road had served as a bridge between the cultures. Upon the road treaded merchants and diplomats, Chinese and Western alike, exporting and importing goods, carrying the Chinese civilization of the magnificent Han and Tang dynasties further and further, deep into the once unreachable lands of Greece and Rome. In 1275, Marco

Polo came to China with his father and uncle. They lived in China for 17 years (1275-1292) and were received by the Emperor of the Yuan Dynasty. In the summer of 1292, Marco Polo, together with his father and uncle, escorted the Mongolian Princess to be married in Persia. Three years later, they returned to Europe. When Polo was kept as prisoner of war by Genoa, he finished the book 'Travels of Marco Polo', wherein he depicted the great prosperity of China, detailing the abundance of silk, condiments, gold, silver and many other treasures, the convenient transportation system and the open and fair social atmosphere. It was an eye-opener for Europeans.

The fifteenth century ushered in the great Age of Sail. In the first three decades, the civilizations from both the eastern and western ends of the Eurasian continent began almost simultaneously to explore the oceans. China's Zheng He cruised far south into the Indian and Atlantic Oceans while Prince Henry of Portugal sailed along the coastline of West Africa. It was one of the peak times in China's history. After two generations of peace and growth under the reign of Ming Dynasty, the country saw a leap in economic development with the formation of its textile center in Songjiang (part of today's Shanghai) and its pottery center in Jingdezheng (the period is also known as the golden age of Chinese pottery).

Zheng He's cruise boosted both the official and private trade of China, exporting textile, pottery and lacquer work, among other art pieces. When Vasco Da Gama reached Cape Town, he learnt from the natives that the Chinese had already been there many times, decades ago. Then the Portuguese fleet, assisted by Arabic navigators, followed the routes of Zheng He to India.

In 1517, Manuel I, King of Portugal, sent Tome Pires as Ambassador for Europe to visit China. Diplomats, merchants, missionaries followed him in swarms, returning with various goods and artworks. Many of them followed in Marco Polo's suit and wrote travelogues and memoirs about China, such as Tractado em que se Cōtam muito per esteso as eousas da China by Gasparda Cruz and China in Ming Dynasty by Martin de Rada [2]. The books helped the Europeans to learn about China, laying foundation for the trend of Chinese style.

In the 17th and 18th centuries, decoration in the Chinese style became popular in Europe. In 1600, Britain established the East India Company, installing a trade station in Sudrajat, west India in 1613, and before long setting up a commercial office in Madras, east India. In 1698, the company bought Calcutta, where the Ganges River flows into the Bay of Bengal, and made its trade headquarters there.

The East India Company's interest was not limited to India's food and industrial materials. It was also attracted by Chinese goods- the company made quarterly purchases in China's Guangdong province. Selling Chinese goods and artworks turned out to be immensely lucrative. Other countries like the Netherlands, Denmark, France, and Switzerland built up their own 'East India Company' equivalents in order to trade with China via India. They imported not only raw materials such as spice, tea, silk and medicine, but also products such as pottery, lacquer work, red porcelain, paintings, wallpaper, sculptures, furniture and screens. The shape and pattern of those artifacts, being so different from the church and Western classic styles, were (and still remain) temptingly exotic to Europeans, luring royal, noble and academic collectors. As the fad for Chinese goods kept growing, the import was trivial compared with demand, leading to the golden age of Chinoiserie in Europe.

1.2 The Glorious Reign of Kangxi and the Sun King Louis XIV

The Kangxi emperor of China and Louis XIV of France were both sagacious kings who accomplished great feats. Both made great contributions to the exchange between Eastern and Western cultures.

In 1667, seven years after inheriting the crown, Kangxi came of age and into power. With unprecedented wisdom and courage, he ruled and served his country diligently, making China the world's largest and most powerful country of the time. He respected diversity and united the many minorities harmoniously. During his rule of 41 years, not a single war was waged and society developed soundly and steadily. It was in the emperor's nature to seek out new things. He kept books close to hand throughout his whole life, not only to inherit from and draw upon traditional Chinese culture, but also to explore Western studies, making himself an expert in mathematics, physics, medicine, chemistry, optics, topography and cartology. He strengthened ties with Western countries such as France and Italy and kept frequent contact with missionaries, bringing their talents to full play in building up a prosperous China and promoting mutual cultural exchanges.

The reign of Louis XIV coincides with China's unprecedented prosperity under the rule of Kangxi. At the time, China's art and culture had reached its zenith with exquisite pottery, calligraphy, painting, furniture and handicraft works of all kinds. In 1697, a French missionary named Joachim Bouvet published the Biography of Emperor Kangxi in Paris, arousing a nation-wide frenzy for China. As China was so powerful and prosperous, its political system, economy, culture, art were all attractive to the French. The Chinoiserie began from the court and nobility. Cardinal Jules Mazarin, prime minister under Louis XIV, served as the most typical example. He held a wide collection of Chinese fans, textiles, cupboards and more[3]. His interest influenced other noblemen and members of the royal family, including Louis XIV himself. The king ordered the Hall of Mirrors in Versailles to be set apart solely for the collection of porcelain. In 1670, Louis XIV built the Grand Trianon for his mistress, Madame de Montespan. The castle assumed a Chinese style with French-made soft porcelain as exterior decoration. It was a fashion landmark of the time. In January 1700, Louis XIV attended a grand ball held in Versailles in a Chinese sedan chair, showcasing his love for Chinese culture. A large amount of the artifacts enlisted in the "Checklist of Court Properties" came from China [4]. In the Queen's rooms at Versailles, Compiègne and Saint Germain stood many sets of Chinese furniture. There were Chinese armchairs and folding chairs in the Prince's office in Fontainebleau, and even in the servants' houses could one find Chinese lacquered tables, porcelain vases and tapestries.

With the increase of direct trade between China and Europe, batches of Chinese artworks landed on the continent and were warmly received by the upper classes. It was the vogue of the time to use Chinese porcelain, lacquered handicrafts and brush painting in décor. The fad reached its peak in the 18th century.

1.3 The Influence of Chinese Artworks

Most Europeans did not have the chance to travel to China themselves, thus their knowledge of the country was mostly based on imported goods. In the 17th and 18th century, artists drew inspiration from Chinese artifacts and brought their impact to countless aspects of life in countries such as France, Germany, Spain, Portugal, the Netherlands and Britain. The Chinese philosophy, namely gentility, elegance, freedom and harmony with culture influenced nearly all aspects of art, including painting, sculpture, clothing, furniture, decoration, architecture, gardening and many more. European design began to feature heavily many unique Chinese elements, including the dragon, phoenix, kylin, tiger, deer, butterfly, bat, plum blossom, orchid, bamboo, chrysanthemum, lotus pond; the three durable plants of winter: pine, bamboo and plum blossom, the peony and the palm. Others included scenery, fables, characters, fights and wars, as well as the flowing contours of water and cloud. It was the Chinese artifacts that brought about a new vision of design: to learn from nature and express one's thoughts and feelings through imagination, breaking the bonds of the rules of perspective, symmetry and proportion. The vision inspired thousands of European artists and, together with the fad for collection, brought the Chinese style onto the front stage of European design and decoration.

Textiles   Chinese silk goods, soft, elegant, elastic and shiny with various delicate patterns, were the first time that linked China with the outside world. The silk goods exported to Europe consisted of greige, or raw fabric, embroidery and hand-painted silk. Silk began its European life as a luxury and gradually found its way to the common houses as a kind of favorite material for clothing, carpets, tapestry, and sometimes wallpaper. Chinese embroidery had a significant influence on the fashions of European royal courts. Embroidery patterns, goffering and bow-knots were common on the clothes of nobles. Some ladies even had Chinese silk, brocade and embroidery on their footwear. Oriental embroidery was most fashionable in the court of Louis XIV. Waliter, a court embroider, and Robin, a court gardener, joined hands in setting up the embroidery association to train embroiderers in Oriental patterns and styles [5]. Hand-painted embroidery became most fashionable in the second half of the 17th century.

Russia is the only European country that shares a border with China, and Chinese style has been an eminent part of Russian art since the era of Peter the Great. Many palaces in St. Petersburg overflowed with Chinese articles and reproductions. The walls of Menshikov Palace were even wrapped with embroidery featuring the Chinese novel, The Soul Hunter [6].

Porcelain   The first written record of Chinese porcelain in European literature dates back to the 5th century. But before the 16th century, the porcelains exported to Europe were very limited in number with extremely expensive price tags. Even in the middle of the 17th century, porcelain was still deemed as a luxury for display and collection in courts, and for home decoration of noblemen. For example, John III, King of Poland, set up a China hall in his palace to display porcelains, and Mary II, Queen of Britain, enjoyed decorating her villa in Poland with pottery. After returning to Britain, she filled her Hampton Palace with porcelain [7].

As China opened its ports for trading and allowed other countries to set commercial offices in Guangzhou, the export of porcelain grew by leaps and bounds. Eventually, it was no longer a luxury and became common dishware for food and beverages. Some were even customized to meet the demands of European merchants. China's production and export of porcelain reached a considerable scale during the Wanli period of the Ming Dynasty, with the European market for porcelain reaching maturity in the late 17th century. In order to better meet demand, manufacturers in Jingdezheng, Fujian and Guangzhou started to accept foreign orders and customize their products in accordance to requirements of shape, size, pattern from the European market. In the book Porcelain Kiln in a Bamboo Grove, Liu Zifen, a scholar of the late Qing Dynasty, wrote, "When the ports were first opened, western merchants had to land in Macau first and then journey to Guangzhou. In the middle of the Qing Dynasty, swarms of ships made berth on the port and the business boomed, as the Europeans loved Chinese porcelain. The Chinese businessmen tried to cater to their pleasures. They produced plain porcelain which was then sent to be painted with Western paintings. Then, the porcelain was roasted into colored porcelain and sold to foreign merchants." The most renowned of such kind are porcelain with coats of arms. Government departments, armies, noblemen and royal families demanded their coat of arms to be painted on the porcelain they ordered. The blue and white porcelain purchased by Manuel I (1469-1521) preserved to this day is recognized as the first porcelain with coat of arms ordered by the Europeans [8].

Rapid as the growth of import was, it could not keep up with the rocketing demand. This led to the birth of porcelain manufacturing in Europe. Early in the end of the 16th century, Arch Duke Francis I supported Italian potters to produce glazed pottery, or Medici, in Florence. A limited portion of their works are held in collection by some European museums. Medici porcelain is very similar to China's blue-and-white porcelain in color and pattern; and is blue and white with patterns of flowers and fruit, in stark contrast to 16th century European colored porcelain featuring biblical themes. As low-temperature ceramics, Medici is markedly inferior to China's blue-and-white porcelain in quality. In the 17th century, many other European cities started to copy Chinese porcelain and purple clay pottery, including Leon in France, Meissen in Germany, Delft in the Netherlands and Staffordshire in Britain; among which, the tin glaze pottery produced in Delft was famous for its similarity to Chinese blue-and-white and polychrome porcelain. The early Delft porcelain was blue patterns on white background, similar to the chinese blue-and-white porcelain, and featured Chinese patterns with dragons, phoenixes, lions, pavilions, mansions, immortals and landscape. It was called by the Europeans as the "image of China" and was widely considered as the most Oriental. Despite its exotic charm, the porcelain was made into shapes fit for daily use in European families, giving birth to a unique style in the European continent. In the 18th century, Meissen in Germany produced porcelain called "Chinese Emperor and his Servants", an immediate success and the most popular product of its time. The images and clothing of the figures on the porcelain however were in fact rather Europeanized, reflecting a mismatch between reality and their concept of Chinese culture [9].

Lacquered furniture and wallpaper   By the end of the 16th century, some lacquered Chinese furniture had already found its way to the houses of rich merchants in Britain and France. Since the time of Louis XIV, the increasingly fashionable Chinese style triggered a craze of copying Chinese lacquer works in Britain, Germany, the Netherlands and other European countries. Black paint and patterns outlined in gold added a magnificence to cupboards, chairs and tables, boxes and screens. The most famous lacquer work craftsmen in Europe were the four Martin brothers from France and Stobwasser from Germany. The British furniture designer Thomas Chippendale published "The Gentleman & Cabinet-Maker's Director" in 1762, which included designs of Palladian architecture, Goethe and Chinese styles. The book played a key role in the spreading of the Chinese style in Europe. Around 1755, Thomas Chippendale made a Chinese style 2.41-meter-wide bed, which now resides in the Victoria and Albert Museum in London. It is a four-column bed with a pointed top. Odd as it is, the bed showcases a real blend of Chinese and Rococo styles [10]. The Chinese traditional landscape painting found on a 1740 Italian drawer chest exemplified the significant influence of Chinese art on European furniture, among which the luxurious gilded pattern wrought the greatest impact. Its influence was not contained to just furniture, but expanded to affect pottery, metal processing, textiles, wallpaper, tapestry and interior decoration.

Wallpapers were an indispensible decoration in the homes of Europeans. The sea powers of the Netherlands and Britain in the 16th and 17th century started to import wallpaper from China, and it was not long before Chinese wallpapers were copied. Wallpapers with figures, birds and flower patterns provided a harmonious match to Chinese furniture, and were thus very popular in Europe, leading to the production of Chinese-style wallpaper in Europe. Designers transposed patterns of Chinese artworks onto wallpapers to create a simple and elegant Oriental exoticism. In 1754, British Jackson and Battersea built up a factory that produced wallpapers featuring Chinese landscape painting [11]. As Chinese style wallpaper became more and more popular, it began to blend with local designs to form the Sino-French and Sino-British styles. The new combinations of luxuriance, simplicity and elegance brought diversity and innovative expressions in Chinese-themed decoration.

Painting   The influence of Chinese painting over European fine art also had its root in the fad for Chinese artworks. Artists featured their works with Chinese objects e.g. Chinese Porcelain Bowl Containing Cherry and Strawberry of 1608 by Osias Beert and Globe and Luxurious Dishware of 1670 by Pieter Borl. Watteau and Boucher, leading figures of the 18th century Rococo school, were both very interested in the Chinese style. The former drew a number of Chinese figures, while the latter designed nine carpet patterns, such as Audience with the Chinese Emperor, Chinese Wedding, Chinese Custom and Chinese Garden, amongst others he also drew a sketch called The Lady Blowing a Musical Pipe, featuring a Chinese lady. Boucher's oil paintings, Chinese Emperor Holding Court, Chinese Garden, Chinese Landscape with Figures, Chinese Fishing [12] and more are collected by the Besancon Museum in France and another museum in Rotterdam. A master of sketching and color, Boucher created a perfect blend of the classic, rococo and oriental styles; precise yet full of amusement and exotic flavor. His works highlighted an elegant, easy and joyful lifestyle.

1.4 The Influence of Chinese Architectural Decoration and Garden Art and Rococo Style

Many aspects contributed to the acceptance and spreading of Chinese architecture and garden art in Europe. Missionaries who had been to China, as represented by Matteo Ricci, were full of praise for Chinese garden design, which transcended form and shape and were thus far more natural and tranquil than the geometrical styles in Europe. Two illustrated books published in Rotterdam helped to spark the Europeans' interest on Chinese garden decoration: A Travel to China with 150 pictures, published in 1655 by Hollander J.Nieuhof, and Chinese Antique Atlas with a detailed introduction of the designing and decoration of Chinese gardens, published in 1655 by Athanasius Kircher. In 1749, the painting Chinese Emperor Walking in his Palace by Jean Denis Attiret was published in the Letters of Society of Christ, introducing Yuanmingyuan, or the Old Summer Palace and Imperial Garden, to the Europeans. Thereafter Attiret mailed copies of Chinese artists' paintings of Forty Sceneries of Yuanmingyuan, the Garden of Prefect Splender (now collected in the National library in Paris) to Paris, which served as drafts [13] for the French to imitate Chinese garden architecture. The European architects also learnt to imitate Chinese garden architecture from images on porcelain, lacquered works and silk.

Curiosity and romance pushed the Europeans toward Chinese architecture and garden art. In many British gardens stood Chinese pavilions, the most famous of which can be found at Kew Garden in south-west London, built by William Chambers for Augustine, the Princess of Wales. The garden features brooks, rockworks, waterfalls, caves, twisted routes and prosperous plants, as well as a nine-story pagoda. Other countries like Germany and the Netherlands admired these gardens and built their own by imitations. In 1781, Landgrave of Kassel built a Chinatown in Weissenstein and named it Moulan. Almost all the buildings in the town were constructed in the Chinese style, and the pagoda and bridge still stand to this day. It became the fashion in Europe to build houses with raised eaves and pointed roofs, and with pagodas and pavilions. The Sans-Souci in Potsdam in Germany and the China Palace on Queen Island in Switzerland boast small Chinese buildings and pavilions [14].

Chinese garden art reflected the philosophy of living in harmony with nature and the pursuit of a tranquil life, which influenced the development of European art and contributed to the birth of the Rococo school and its diversity as well as inclination to nature. Some believe that the delicate, exquisite and luxuriant Rococo artworks were derived from the Italian Baroque style. Others hold that they stemmed from the brooks and rockworks of Chinese garden art. Nonetheless, it is certain that Chinese garden architecture nourished and nurtured the development of the Rococo style [15], characterized by the application of curly and flowing lines and the use of drawings of nature. Japanese scholar Kobayashi even calls Rococo the Sino-French style. The influence of the Chinese style was not limited to France, but simmered into every aspect of European culture, with the great geographical discoveries and the establishment of the global market. Even the rise of Art Nouveau in the late 19th century was mainly affected by the Chinese philosophy of harmony with nature, only with different forms and perspectives. This was the period when the legend of Cartier began. The legacy of Chinese style and Art Nouveau undoubtedly had a significant influence on the understanding of fashion and beauty of the avant-garde designers of the brand.

2. The Chinese Elements in Cartier Jewellery

The story of Cartier began in 1847 when Louis-François Cartier, then aged 28, took over his teacher's jewellery workshop at No.29 Montorgueil Street, Paris, more than one hundred and sixty years ago. A year earlier, while he was just a craftsman, Cartier designed the heart and diamond shaped trademark out of his initials L and C and had it registered [16]; evidence of his remarkable business talent. The Cartier family read extensively on history and collection of art. They also travelled around the world to collect antiques and artworks from Egypt, China, Japan, India, Russia, Arabia and Persia. In Cartier's historical records, there is a picture of Mademoiselle Jeanne Toussaint, once a Senior Jewellery Director of Cartier. The picture was taken in 1925 [17].

This elegant, fashionable lady sat charmingly before a lacquered Chinese screen from the middle Qing Dynasty. The decoration and painting of the screen had not only survived the ages but still shone with brilliance.

Cartier's features, with their deep cultural tradition and sublime elegance, are a result of the family's travels around the world, their comprehensive collection and extensive learning. For generations, designers took inspiration and innovation from foreign art, and strive to not only imitate but also to transcend and go beyond. They define the greatest luxury and elegance with divine combinations of gold, silver and precious gems.

The Cartier jewellery of the 1920s and 1930s clearly indicates how the brand was affected by the Chinese decoration style that crossed over from the 19th century. The designer seemed captivated by Chinese art and tried very hard to integrate Chinese elements into his works; the strong Oriental flavor of which captivated the world. The influence came mainly from jade carring, lacquer works and silk textiles. The jewellery had defining features as follows:

2.1 Directly Borrowed Chinese Artworks

Some Cartier jewellery are based on existing Chinese artworks, adopting the same materials, shapes and patterns while adding in some decoration of gems to make them more fitted to the Western eye. Being full of Oriental flavor and in line with Western taste, these baubles became one of the most characterized collections. In 1912, Cartier made a candy box (BS 05 A12) out of a jade snuff box from the Qing Dynasty by adding a buckle decorated with round and olive-shaped, cabochon rubies and rose-cut diamonds. This is one of Cartier's earliest works in the Chinese style.

In 1921, Cartier Paris made a pendant (PE 15 A21), based on a top-quality piece of jadeite with a lotus pattern from the Qing Dynasty. Rubies and rose-cut diamonds were added to cater to the European taste. Luscious red combined with jadeite green is a perfect example of the traditional Chinese color combination. In 1924, Cartier Paris designed a brooch (CL 80 A24) full of mystery from a jade dragon belt-hook with sapphire, diamond and black enamel decoration. In 1935, Cartier Paris placed a obsidian bottom under a lotus brush washer and transformed it into a delicate ashtray (SA 04 A35) .

The above articles from Cartier Paris were all made directly from existing antiques. The designer aimed to maintain the original flavor with little ornaments as the bridge between cultures. In 1928, Cartier Paris made a snuff box (PB 20 A28) with two open carved jade on the upper and lower sides. At the center of the cap was a gold twist that joined with a lapis lazuli ring. The sides of the boxes were covered by cloisonné with green and blue patterns. In 1930, Cartier London linked 21 pieces of round jade together with gold rings to form a belt (JA 26 A30). Each piece of jade was made into the shape of coins during the Qing Dynasty and carved with the characters " 光绪通宝 "(the currency of Emperor Guangxu). Apparently, the design stemmed from the jade belt of the Ming Dynasty, showcasing Chinese tradition and decoration to a unique effect.

Another type of Cartier jewellery is of typical European style with some Chinese elements. Western as they are by nature, the Oriental components highlight this exotic hybrid flavor.

Take the vanity case(VC 14 A26)for example, designed into the shape of a Chinese pillow covered with black enamel and with gilded patterns of " 卍 ". On the two sides of the pillow are two delicately carved stork beak plaques of five bats and the character " 寿 "(long life), as an omen for happiness and longevity.

The pair of plaques were formerly independent articles and where the inspiration of this magnificent vanity case came from.

The crouching dog paperweight (DI 08 C27) is made mostly of dark jade; on its top crouches a divine dog, which had previously been an independent ivory sculpture of an auspicious animal from the late Qing Dynasty. This piece of jewellery combines ivory with dark jade, accompanied with sapphire beads and a red enamel cloud silhouette, harmonious and mysterious.

The buckle of the belt (JA 05 A28) is a Chinese jade board embedded with turquoise, sapphire and a blue silk band. It remains fashionable even in our time.

The seal watch-brooch (WB 34 A29) is composed of gold, platinum, diamond, ruby beads, agate and enamel. It is attached to a 19th century jade lion-shaped seal, at the bottom of which is embedded a watch.

In addition to ornaments, Cartier also boasts expertise in making premium watches and clocks. The use of Chinese artworks has led to the birth of many classic Cartier watches and clocks. Among them is the carp clock (CS 11 A25).Based on a pair of jade carps with a sector fly-back clock on top and an obsidian bottom, together with other detail decorations of gold, platinum, coral, pearl and diamond, it is a clock with distinctive Chinese flavor and authentic innovation.

The mysterious chimera clock (CM 23 A26)brings the Chinese divine animal onto a clock made of gold, platinum, pearl, diamond, enamel, yellow crystal (dial plate) and red coral. It is not only extravagant and exquisite but also mythical and marvelous. The inspiration for it came from clock design from the time of Louis XV and Louis XVI, when mechanisms were commonly built into the back of the animal. It is one of Cartier's landmark collections.

The mysterious goddess clock (CM 04 A31) is based on a 19th century white jade sculpture of Kwan-yin. On the sides of the Goddess lie a white jade auspicious animal and a blue jade rush pot. The coral in the pot is embedded with pearls and turquoise. In front of Kwan-yin is an octagon-shaped clock with transparent crystal dial plate and turquoise edging. The jade Kwan-yin itself was already a jade masterpiece, and the Cartier designer's reworking brought dimension and congenial colors together for a better visual impact. Kwan-yin saves the masses from suffering and the auspicious animal brings peace and luck- this is a true showcasing of the supplicated Chinese culture.

The turtle shaped chimera clock (CS 05 A43) reused a 19th century red coral turtle carrying a bottle dismantled from another clock dated 1928. It has a white onyx base and clock dial, diamond hands and scale. The red coral turtle is also embedded with diamond ornaments. The clock casts an impression of tender brightness. The combination of turtle and clock symbolize peace and long life in Chinese culture.

Some of the Chinese-style Cartier jewels are processed from parts of original items. The screen base clock(CDB 23 A26) is one of them. The dial plate is clearly made from part of a jade relief screen from the middle Qing Dynasty; partly wrapped with black enamel edging embedded with red coral. The clock, with European hands and detailed finishing of gold, platinum, diamond, pearl and enamel, is an impeccable piece of art that highlights Chinese culture and the sublime elegance of Cartier.

The palm leaf click pin(CL 244 A25) is made from the remains of jade worn in the Qing Dynasty. Words of good omen can still be seen on the jade pieces: " 福在眼前 " (Happiness is right in front of us) [18]. Combined with ruby, diamond, gold, black enamel, the remarkable talent of the designer has turned remains into an exquisite artwork of Indian style and Chinese flavor.

2.2 Absorbing Chinese Elements

Absorbing and integrating essence of various cultures into its masterpieces is another of Cartier's defining features. A large portion of Cartier jewellery draws upon Chinese elements of painting, pottery, lacquer, silk, jade, and enhances their implications. The genteel quality of Chinese culture blends perfectly with the sublime elegance of Cartier, bringing to its latest fashion a touch of age-old heritage.

The pendant of the *yin-and-yang* (NE 19 A19) was made by Cartier Paris in 1919. It applied the Taichi diagram on a pair of round pendants made of black enamel and diamond and embedded ruby and emerald ornaments, forming a new expression of the philosophy that Taichi is the source of the universe.

The " 寿 "(long life) shaped earrings (EG 24 A26) and the Maitreya earrings (EG 11 A28) feature two important Chinese cultural elements in their design. The character " 寿 " has a long history of being used as a decorative element, as the many variants of the character are constantly used to symbolize long life and good luck. Maitreya is a Buddhist deity; it is believed that his packet can contain disasters and take them away.

In its early works, Cartier used several dragon shapes and patterns and formed the Chimera series. In 1923, Cartier Paris made a Chimera click pin (CL 260 A23) from the shape of a pearl-playing dragon, and the dragon-shaped bracelet has even become a classic in the Cartier collection. The Chimera bracelet (BT 109 A28) features two dragons playing with one burning ball. The heads are carved from coral and match well with the green emerald beads, using the classic Chinese colors. It also integrates many other precious materials such as gold, platinum, diamonds, sapphires, blue, black and green enamel and more, with extremely delicate design.

The platinum chimera bracelet (BT 064 A29) is the first all-platinum gem chimera bracelet from Cartier, made of platinum, diamonds, sapphires and transparent crystal. The heads of both dragons are identical to typical Chinese dragons and join with each other on the front.

Cartier Paris also launched two products that feature the dragon and tiger patterns. The brooch (CL 258 A22) has a tiger pattern formed of embedded diamonds on a black enamel ring, reflecting the influence of the tiger patterns on

Chinese jade and bronze antiques. The other is the Cartier vanity case with a dragon pattern (VC 69 A27). The dragon pattern it adopted is commonly seen on the textiles of the late Qing Dynasty.

The inspiration of the pendant (NE 03 A22)came from the Shang and Zhou Dynasty. On its red coral column is embedded a vivid diamond dragon pattern, simple and delicate. In the Zhou Dynasty, there was a significant improvement in the art of jade carving. Gems of different stiffness such as crystal, agate, turquoise and jade can be combined and matched at the craftsman's will, allowing for greater diversity in color and shape. The most typical of these were stringed ornaments and jade wear excavated from the tomb in Shanxi's Wo County and Henan's Sanmenxia. They are mostly semi-circular jade ornaments with small accessories of jade beads, tubes and rings in a balanced structure and congenial match. The group jade wear excavated from Shanxi consists of various forms and shapes of ornaments, including beads, tubes, semi-circular, rings, fang-shapes and others, with all kinds of precious stones such as jade, agate, turquoise and refined coal. It was delicate and extravagant and not bound by any particular style. Jade collarets, wristlets, stringed ornaments and group jade wear unearthed from Henan are similar to those excavated from Shanxi and tell of the development of pendant making of the time.

Some Cartier creations are based on the shape of the Chinese flower basket and vase, including two brooches made by Cartier New York and Cartier Paris (CL 47 A27, CL 48 A28) and two vanity cases (VC 68 A27, VC 71 A28). They were built on the Chinese flower basket and vase and feature a strong Chinese flavor.

The vanity case (VC 67 A24) is made of two wooden boards and many other insets and carvings of various materials, telling a Chinese legend. Two armed heroes greet each other. Around them is a herd of sheep, and on their heads a flock of cranes flies through auspicious clouds, a scene of tranquility and peace. There are two characters of "寿" carved in diamond, up and down the case on the turquoise plates. The case is topped with a diamond sculpture of a ball-playing dragon. All its features are typical and meaningful Chinese elements. The main image was made via a process similar to the making of a Chinese treasure chest with various gems including red coral, turquoise, pearl button, lapis lazuli, malachite and agate.

The ink bottle (DI 07 C27) was made by Cartier in 1927. It builds on the shape of a Chinese incense burner. The cap is carved into a traditional Chinese pattern of branches and flowers. The joints between the wood base, the body and the cap are typical Chinese elements.

The vanity case (VC 72 A28) made by Cartier Paris in 1928 is another good example of how the brand integrated Chinese elements in its designs. The main image is from a polychrome porcelain pattern collected by Louis Cartier [19]. It is a typical Chinese lady painting depicting an elegant lady sitting at a stone table in the corner of a garden. The plum flower in the vase and the pine and bamboo behind her are the three durable plants of winter, embodying the quality of perseverance.

2.3 Materials and Processing technique with Typical Chinese Characters

Cartier exercises great care in choosing materials. The designer pays much attention to the relationships between materials; material and shape, material and color. The layered and congenial matching of gold, platinum, diamond and other gems highlights the designer's creativity and aesthetic achievement. Some gems are most common in the works that merge the charm of Western and Oriental cultures; namely emerald, jade, coral, agate, crystal and turquoise which are also frequently used in the traditional Chinese handcrafts. China was the first country to apply jade, crystal, and emerald in jewellery, dating back to the Neolithic period. Jade is one of the favorite stones of the

Chinese. The most pure and tender of these were selected for rituals in early times and secular life in the more modern ages. The charm of jade comes more from the cultural connotation it bears than its natural beauty. It has been linked to morality, power and deity, forming an enduring jade culture of many a millennia.

In the broad sense, crystal, agate, emerald and turquoise all belong to the jade family. They play a vital role in China's art of decoration and ornaments.

Crystal is a transparent variety of the silica mineral quartz that is valued for its clarity and total lack of color or flaws. Quality crystals are beautifully glistening hexagonal prisms closely linked to each other. When mixed with different minerals during formation, they can assume various glittering and translucent colors.

Agate is a kind of common semiprecious silica mineral, a variety of chalcedony that occurs in bands of varying color and transparency. It is hard and smooth with various shapes and colors, suitable for the processing of industrial art.

Turquoise is a hydrated copper and aluminum phosphate that is extensively used as a gemstone. Different element compositions lead to various colors. Its most characteristic color is azure.

Being the skeleton of marine organisms, coral is special amongst the other gems. Held dear to the Tibetan Chinese who carry them as amulet and a way to communicate with the heavens, they are commonly found in stringed ornaments from Neolithic times and the Shang and Zhou Dynasties.

Jadeite is a grass-green variety of beryl that is highly valued as a gemstone. It is glistening, hard but not crisp and has a color pleasing to the eye. Being recognized as the King of Jades, it has been widely used as raw material for ornaments and was most popular in the Qing Dynasty. As the emerald was the favorite stone of emperors and noblemen, especially Emperor Qianlong and Empress Dowager Cixi, its priced rocketed and was nicknamed as "the imperial jade".

The western craftsmanship is best known for the making of pottery, wood, glass and metal. Cartier jewellery, however, combines gold, platinum, diamond rather freely with various other gems, demonstrating its expertise in all precious metals and stones. Combined with exquisite workmanship, Cartier has long excelled in bringing the beauty of the gems into full play.

The clock (CDB 07 A27) made by Cartier Paris in 1927 serves as a good example of Cartier's brilliant interpretation of the beauty of agate. The agate body of the clock is carved into a Chinese traditional pattern of a bird perched on an abloom tree in the handle-shape of Qing Dynasty fashion. The dial plate and base is made of lapis lazuli, while the clock face is made of yellow crystal. Lapis lazuli, agate and yellow crystal form a contrast of color. The octagon clock face and the square base sets off the shape of the agate body. The clock carries a strong Chinese flavor, natural colors and diversified yet congenial shapes.

There is one type of Cartier products that are made in a process very similar to the Chinese traditional technique of inlaying mother-of-pearl, used in lacquer works. Shells of various colors are polished into various shapes and sizes and put into pre-designed patterns. Then the bits of shells are painted and polished until they become an integral and colorful part of the lacquer. The technique was first developed in Western Zhou Dynasty and came to maturity in the Tang Dynasty. It reached its peak in the Ming and Qing Dynasties, with the prosperity of lacquer ware and furniture manufacturing. Due to their magnificent appearance, lacquer wares inlaid with mother-of-pearl enjoyed great popularity in the courts and among the masses. The technique is applied in many parts of China including Beijing, Shanxi, Guangzhou, Suzhou and especially Yangzhou [20]. The technique can be divided into the soft method and the hard method, each leading to different effects. Patterns made with it are widely applied in various daily objects from doors and windows to furniture, from dishware to stationary and more.

Some Cartier artworks also have mother-of-pearl laid in red or black enamel base, very similar to the Chinese technique. The traditional Chinese patterns are retained with more fashionable designs and luxurious ornaments. They came in two forms.

Some used mother-of-pearl to decorate a clock's base, background or dial. The phoenix powder box (VC 58 A25) and dragon powder box (PB 21 A27) both have patterns inlaid with mother-of-pearl as a base on which the phoenix or dragon are put together with golden and black enamel. The two boxes, brimming with golden lines and cloud-shaped patterns, are typical works featuring Chinese style. Both the dials of the clock base (CBD 05 A26) and clock with crouching dog (CBD 12 A26) are inlaid with mother-of-pearl, magnificently glistening and glittering .

The others used inlaid mother-of-pearl to make patterns for decoration. The auspicious cloud pattern at the center of the vanity case (VC 54 A25) was put together with inlaid mother-of-pearl. On the cigar case (TB 07 A22) inlaid mother-of-pearl formed a Chinese landscape. The surface of the cigar case (TB 11 A25) is designed into a scroll of a Chinese painting; on it are landscapes and figures put together by mother-of-pearl. On the four sides of the base of a repeater clock (CR 17 A30) are figures and time scale made by inlaid mother-of-pearls, adding luminous colors to the solemn clock. The jade column and animal button on its edges make it even more Chinese.

2.4 Snuff Bottle – a Typical Combination of Eastern and Western Cultures

Several Cartier Chinese style products with special characteristics were built on snuff bottles.

Snuff was first brought to China by missionaries. When snuff first appeared in Kangxi's reign (1622-1722), it was contained in boxes and was very volatile. To make it easier to carry and preserve, snuff bottles in various shapes and materials came to the market. They were first made by court manufacturers and soon became popular amongst the people. Snuff bottles came to China at a thriving time when the many branches of craftsmanship [21] reached their peak, and integrated the various arts of pottery, glass, jade carving, enamel, crystal, gems, the carving of bamboo, wood, bone and horn and interior painting. Soon it transcended from an object of daily use to a kind of all-rounded artwork, an example of the merging of Eastern and Western cultures. The snuff bottles of jade and coral were at their most magnificence in the Qing Dynasty. With little modification, Cartier designers turned them into exquisite and luxurious perfume bottles and lighters. The body of the perfume bottle (FK 06 A25) is a snuff bottle of pure and glossy green jade with a natural texture. Its cap and base are made of blue and black enamel, gilded with gold and embedded with rubies. The body of the perfume bottle (FK 11 A25) is a snuff bottle of flat round jade with a fine texture and bright pure color, and embedded with sapphire beads and the character" 寿 "(long life). Its cap and base are made of blue and black enamel, gilded and embedded with round sapphires. The addition of gold and enamel turned these beautiful bottles extravagant.

The body of the perfume bottle (FK 05 A26) is a yellow jade snuff bottle with wide shoulders and a large belly. The jade is smooth and has a leveled shade. Its gilded fan-shaped lapis lazuli cap forms an elegant contrast of color with the yellow body. The body of the perfume bottle (FK 19 A26) is a coral snuff bottle with lotus pattern. The Cartier designer kept the original shape of the snuff bottle and added only very few diamonds, gold, pearl and black enamel ornaments. The table lighter (LR 22 A39) was built upon a white jade snuff bottle of a pure and tender shade and carved scriptures. On the top of it lies a gilded black enamel cap with coral and diamonds embedded.

The design of Cartier follows the trends of fashion. These works of strong Chinese flavor are results of the Chinese style popular in Europe before the 20th century, especially Europeans' adoration of Chinese culture in the 18th and 19th centuries. The Age of Sail opened the door of Chinese civilization to the Western world, allowing Cartier designers to draw inspiration from Chinese artworks. With great expertise, perfect skills and unparalleled imagination, they showed the world their understanding of and fascination with the Chinese culture, with precious masterpieces of art.

3. Conclusion

This is wisdom and skill behind fashion. The Chinese style Cartier artworks revealed to us the immersion and influence of Chinese culture to the Western art of ornament and decoration. Talented designers showed great eagerness in absorbing the artistic essences of other cultures, either by directly borrowing the elements or integrating selected ones. Their work and spirit transcended the limits of time and space and, through constant innovation and faithful adherence to the highest standards, contributed to the world enduring beauty and romance for over sixteen decades. Each piece of Cartier jewellery has its own character, yet are all unified in their style of simplicity, fashionability and elegance. Cartier, full of culture and elegance, has become a legend that outlives fashion and brings inspiration to generations.

the Palace Museum

Song Haiyang

Notes:

1 2 *Chinese Style Designing in Europe between the 17th and 18th Centuries*, Yuan Xuanping, Antique Press, 2006

3 10 12 *European Fine Art, from Rococo to Romanticism*, Yuan Baolin et al, China Renmin University Press, 2004

4 *Louis XIV, le Roi Soleil : Tresors du chateau de Versailles,* edited by the Palace Museum and the Versailles, Forbidden City Publishing House, 2005

5 *Oriental Silk Art in the Western Eye*, Ma Liang, Shanghai Education Press, 2004

6 *Chinese Elements in St. Petersburg*, Han Xianyang, Guangming Daily, Sept. 5, 2007

7 9 11 14 *Missionaries in the Ming and Qing dynasties and European Sinology*, Zhang Guogang, China Social Sciences Press, 2001

8 *Oriental Pottery Art in the Western Eye*, Chen Wei and Zhou Wenji, Shanghai Education Press, 2004

13 *History of Cultural Exchanges between China and the World*, Zhou Yiliang, Henan People's Press, 1987

15 *Foreign Garden Art*, Chen Zhihua, Henan Science and Technology Press, 2001

16 *Cartier*, Cartier

17 *Treasures of Cartier*, Shanghai Museum, Shanghai Fine Arts Publishing House, 2004

18 Christie's Hong Kong, Sales catalogue of Nov. 6, 1977

19 No. 101 record of auction Nov. 25-27, 1979 Monaco Sotheby's provided by Alain Cartier

20 *History of Mother-of-Pearl*, Shen Congwen et al, Wanjuan Press, 2005

21 Snuff Bottles, the Palace Museum, Shanghai Science and Technology Press, 2002

References:

History of Sino-foreign Relations Volume 4, Shanghai Translation Publishing House, 1988

Exchanges between Eastern and Western Fine Arts, Michael Sullivan, Jiangsu Fine Art Press, 1998

Chinoiserie in 18th Century Europe, Xu Minglong, Shanxi Education Press, 1999

The Influence of Oriental Aesthetics on the West, Chen Wei et al. Xuelin Press, 1999

China and Japan in World Fine Art, Liu Xiaolu, Guangxi Fine Art Press, 2001

The Spread and Impact of Chinese Culture in the 18th century Western Europe, Yan Jianqiang, Publishing House of Central Academy of Fine Arts, 2002

The Export of Chinese Porcelain and the Production of Export Porcelain, Li Huibing, *China and Sweden : Treasured Memories*, Forbidden City Publishing House, 2005

Art History on Foreign Craftsmanship, Zhang Fuye, Higher Education Publishing House, 2006

皇帝的珠宝商　珠宝商的皇帝

作为珠宝史上的先驱，卡地亚始终引领未来风潮，掌握全新工艺，追寻敏锐直觉，创造出独一无二的经典风格。

作为卓有远见的珠宝商，卡地亚自创立至今，一直密切关注社会风尚的更迭演化，并以其珠宝作品再现这些非凡变迁。历史、经济、社会，从来都不会停下前进的脚步。正是它们的持续演进，成为滋养卡地亚创造力的灵感源泉。

图 1　路易·弗朗索瓦·卡地亚
1847 年，路易·弗朗索瓦·卡地亚从其师傅阿道夫·皮卡手中接管位于巴黎蒙特吉尔街 29 号的珠宝店。
Cartier Archives © Cartier

从圣厄斯塔什到皇室宫廷

1847 年，路易·弗朗索瓦·卡地亚（1819～1904 年）（图1）从其师傅阿道夫·皮卡手中接管位于巴黎圣厄斯塔什区蒙特吉尔街 29 号的珠宝店，卡地亚的经典故事就此展开。

依照当时的惯例，卡地亚珠宝店最初只售卖其他珠宝工作坊的作品。

1852 年，路易·弗朗索瓦·卡地亚将珠宝店迁至小场街 5 号，位于时尚的皇宫区后街及富丽堂皇的奥尔良宫殿附近。他的第一位贵宾是拿破仑三世时期的一位重要艺术人物——乌韦克尔克伯爵夫人。当时正处于七月王朝时期，时尚风标仍然为中产阶级的品位所左右。贝壳浮雕风行一时，伯爵夫人在卡地亚的第一份订单，就是一条复古式贝壳浮雕项链。

乌韦克尔克伯爵是拿破仑三世的御前美术总监。通过伯爵夫人的引荐，卡地亚于 1856 年获得了玛蒂尔德公主（图2）的第一份订单。三年后，欧珍妮皇后亲自从卡地亚订制了一套纯银茶具。

此时，位于小场街的珠宝店对于已成为皇室供货商的卡地亚而言太过狭小。因此，卡地亚再次迁址至意大利大道（图3）。此时正是奥斯曼改造巴黎的前夕，意大利大道已经呈现出熙攘繁华的风貌。淑女们身着蓬裙来来往往，人群潮水似的涌入咖啡馆，为这条大街带来络绎不绝的客流。第二帝国时期巴黎上流社会的所有精英人物都汇聚到了卡地亚精品店。

1874 年，阿尔弗雷德·卡地亚（图4）开始执掌家族生意，而此时的境况却大不如前：法国王室流亡到了英国，巴黎公社改变了整座城市的生活。

尽管经济和政治气候变幻莫测，卡地亚仍然保持了多样化作品的生意，包括扇子、韦奇伍德徽章、珠宝、链表、怀表等。从精品店的库存登记簿（图5）可知，这些作品囊括了当时正流行的古典风格珠宝和工艺品，以及一系列传统钻饰。

卡地亚的客户逐渐增加。店铺的账簿记录显示，除法国贵族家

图 2　玛蒂尔德公主
玛蒂尔德公主（1820～1904 年），拿破仑一世侄女，于 1859 年将卡地亚引荐给欧珍妮皇后。
© Collection Roger-Viollet

图 3　意大利大道
© Roger-Viollet

图 4　阿尔弗雷德·卡地亚
阿尔弗雷德·卡地亚 (1841 ～ 1925 年)，
路易·弗朗索瓦·卡地亚之子。从 1874 年
起执掌家族生意。
Cartier Archives © Cartier

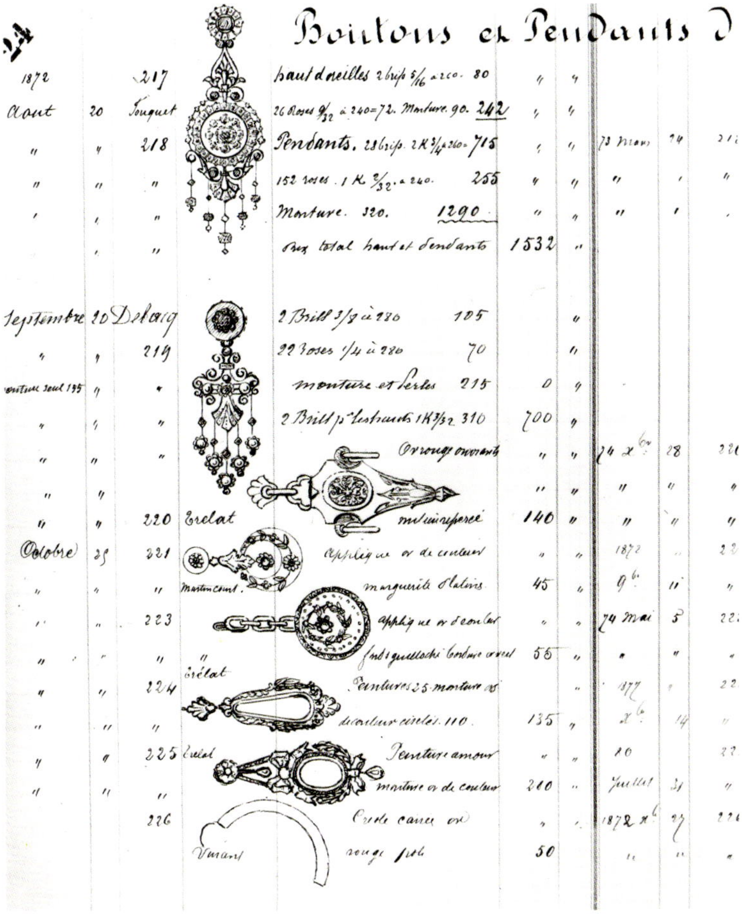

图 5 卡地亚库存登记薄
Cartier Archives © Cartier

庭外，销售名单里也出现了越来越多的银行家、实业家和其他富有的商界人士。

卡地亚与传统贵族和富裕的中产阶级均建立了特殊的关系，后者的一掷千金更是彰显出他们想为自己佩上权力象征的强烈愿望。

这些新兴的富豪阶层更具国际性。卡地亚精品店再次搬迁至高贵优雅的和平街，为的就是满足这些国际客户的期望。这个高尚街区也确实汇聚了众多时尚工艺师：珠宝商、裁缝师、制帽商、手套商、制鞋商和香水商。

和平街与花环风格

迁至位于法国高级珠宝中心的和平街 13 号 (图6)，推动了卡地亚从珠宝销售商到制造商的转型。1898 年，卡地亚成立了一间设计工作坊。1899 年 11 月，卡地亚新店正式开业。就在这两大盛事接踵而至的同时，花环风格的新型铂金珠宝也悄然出现。

阿尔弗雷德将长子路易（1875 ～ 1942 年）作为合作伙伴进入公司。路易·卡地亚 (图7) 挚爱古典文化，热衷于收集 18 世纪的家具和珍玩。他对珠宝拥有自己的独到见解，并有意识地扬弃了新艺术派风格。

卡地亚采用铂金为基座。这种金属更加坚固明亮，但却很难驾驭。铂金的使用推动了宝石镶嵌工艺的变革。除却全新的工具和金属铸造方法外，这种工艺还为珠宝赋予了一种自然流畅的褶皱造型。铂金更易于进行灵动的直线和曲线造型，而此前所常用的纯银基座则显得过分笨重。银的另一个劣势是会削弱钻石的光芒。现在，在当时刚刚诞生的电灯光线下，宝石很难像在摇曳的烛光下那样，散发出璀璨闪烁的光影。

蕾丝、铁艺、花格等表现手法，为花环风格创造出透明而轻灵的特质。铂金为珠宝赋予了生动的造型，以别致的花卉设计、叶形

图 6 和平街 13 号卡地亚精品店
Cartier Archives © Cartier

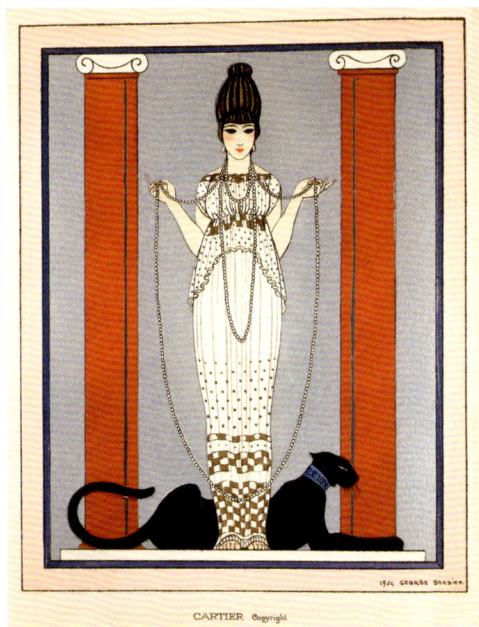

装饰、花环、桂冠、蝴蝶结和滴露形状,大胆地诠释出各种古典和
新古典风格的图案。

　　继中产阶级品位、新文艺复兴以及新巴洛克风格的影响之后,
路易十六风格又再度复苏。这也得益于传统贵族和新兴社会精英所
倡导的古典品位。精致的内部装饰,奢华的舞会场景,为贵族与新
兴富裕阶级的聚会赋予了一种令人追忆的盛典氛围 (图8)。

图9 英国国王爱德华七世与亚历珊德拉皇后
Cartier Archives © Cartier

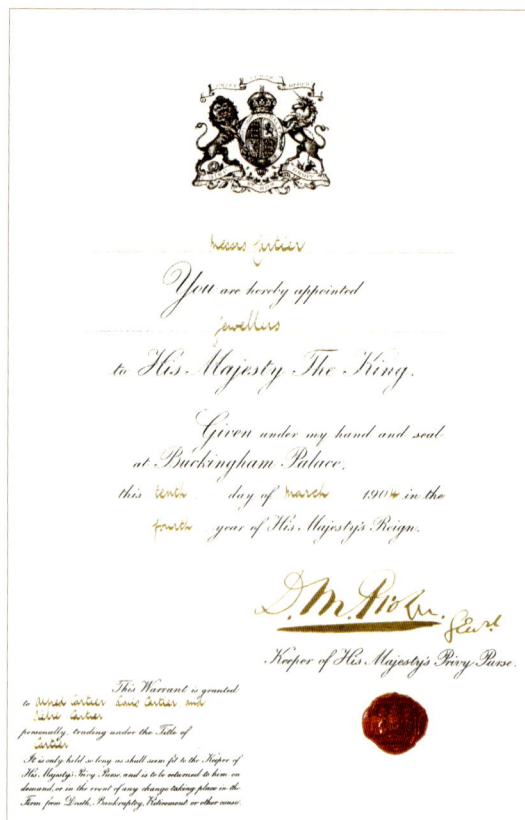

图10 英国国王爱德华七世授予卡地亚的皇家委任状
Cartier Archives © Cartier

钟情于古典品位的新客户

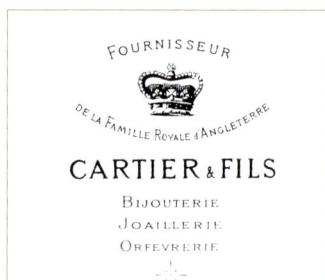

图11 卡地亚信纸抬头
1904年，收到英国国王爱德华七世授予的皇家委任状后，卡地亚使用的信纸抬头。
Cartier Archives © Cartier

伴随工业革命而来的"美好年代"，是一个盛况空前的成长时期。

随着交通和通信工具的现代化，以及殖民地财富的大量涌入欧洲，工业巨子和银行家也以因为冒险精神和机缘巧合而迅速积累了大量财富。

尽管缺乏贵族血统，他们的财富却足以将他们带入最顶级的社交圈。现在，他们是君主们所信赖的顾问，他们的生活方式和行为举止也均以贵族为楷模。

在与这个新兴精英阶层建立良好关系的同时，卡地亚也并未忽视欧洲及更遥远国度的宫廷客户。在1904~1939年期间，卡地亚共

图 12 玛丽·波拿巴
玛丽·波拿巴，嫁给希腊国王乔治一世次子。图为其佩戴着由希腊乔治王子委托卡地亚为她制作的铂金钻石橄榄叶冕冠。
Cartier Archives © Cartier

图 13 卡地亚巴黎为玛丽·波拿巴的婚礼做的展示
1907 年波拿巴公主授予卡地亚在其和平街 13 号店橱窗展示她数件结婚首饰。
Cartier Archives © Cartier

获得了 15 个皇家委任状，成为不同皇室家庭的御用珠宝商。卡地亚的第一份皇家委任状由英国国王爱德华七世于 1904 年授予 (图9、图10、图11)。此后，卡地亚又获得了更多王室的认可，其中包括西班牙阿方索十三世、沙皇尼古拉二世、希腊国王乔治一世、罗马尼亚玛丽王后以及暹罗国王。

拥有一件卡地亚珠宝，即意味着进入了一个特殊的社交领域。在这里，要保持自己的地位，既需要效法和模仿，也需要不断地寻求卓著与差异。因此，卡地亚为皇室新娘提供独特的"婚庆花嫁"。从玛丽·波拿巴 (图12、图13) 奢华铺张的婚礼花篮，就可窥见这种上流社会标志的偏好；从各大工作坊为新娘所制作的妆奁之中，也可明显看到同样的品位。吕西安·波拿巴的这位杰出后裔将皇室风格发挥到了极致，从卡地亚为她所创作的传承拿破仑家族精神的胸饰、肩饰和橄榄叶冠饰就可见一斑。

这些作品均为典型的花环风格。正是这一风格为卡地亚奠定了基础，带来了其发展所需的国际知名度。

图14 阿尔弗雷德·卡地亚与其三个儿子
从左至右：皮埃尔掌管卡地亚纽约业务；路易掌管卡
地亚巴黎业务；雅克掌管卡地亚伦敦业务。
1922年，拍摄于圣让德吕兹。
Cartier Archives © Cartier

图15 1909年的卡地亚广告
广告中提到了卡地亚巴黎、伦敦及纽约的分店地址。
Cartier Archives © Cartier

图16 英国亚历珊德拉皇后
在画中，皇后佩戴着卡地亚"resille"钻石颈饰。该颈饰
由白金汉宫于1904年委托皮埃尔·卡地亚为其制作。
肖像画作者：弗朗索瓦·弗拉孟。
© The Royal Collection, Windsor

图 17 雅克·卡地亚
掌管卡地亚伦敦分店。
Cartier Archives © Cartier

图 18 1929 年的卡地亚伦敦分店
Cartier Archives © Cartier

卡地亚三兄弟与国际拓展 (图14、图15)

1901 年 1 月 22 日，继位近 64 年的维多利亚女王驾崩。在皇室哀悼期结束之后，英国第一家庭即兴致盎然地筹备爱德华七世的加冕典礼。当时的英国仍然是一个保守的国度，卡地亚伦敦分部所创作的珠宝迅速引起轰动。一时之间，冠饰、项链和胸饰订单纷沓而至。由于业务量太大，阿尔弗雷德·卡地亚不得不委派次子皮埃尔（1878～1965 年）及其代理人阿尔弗雷德·布森前往伦敦，以确保这些来自英国贵族的委托订单能获得完美品质 (图16)。

卡地亚伦敦分店于 1902 年设立。四年后，阿尔弗雷德·卡地亚的幼子雅克（1884～1942 年）执掌位于新庞德街 175—176 号（1909年，从新伯灵顿街 4 号搬迁至此）的卡地亚伦敦分店，并从此长驻于此 (图17、图18)。

伦敦分店一开张，卡地亚就收到了来自美国的召唤。纽约当时汇集了三百多位百万富翁，拥有大量新兴财富。无疑，正如摩根、范德比尔特、利兹家族都拥有丰厚的资产，这些财富将他们引入了一个全新的社会阶层。

从第五大道的豪宅，到大都会歌剧院的包厢，不甘示弱的上流社会女主人竞相炫耀自己的显赫特权及豪奢花销，名贵宝石正是她们最钟爱的武器（图19）。

1909 年，卡地亚开设美国分店，赢得了新世界的社会和经济精英的热爱，而他们早已是巴黎珠宝店的老主顾。皮埃尔·卡地亚（图20）被委以重任，在第五大道第 712 号开启了卡地亚纽约的未来。1917 年，卡地亚将纽约业务迁至位于第五大道 653 号的一座文艺复兴风格的大厦内。这座大厦自莫顿·普朗特手中换得。这是一笔非同寻常的交易，整座建筑乃是以一条精美的两串式珍珠项链象征性交换而来，这两串珍珠分别有 55 颗和 73 颗，深受莫顿·普朗特的年轻妻子之青睐。

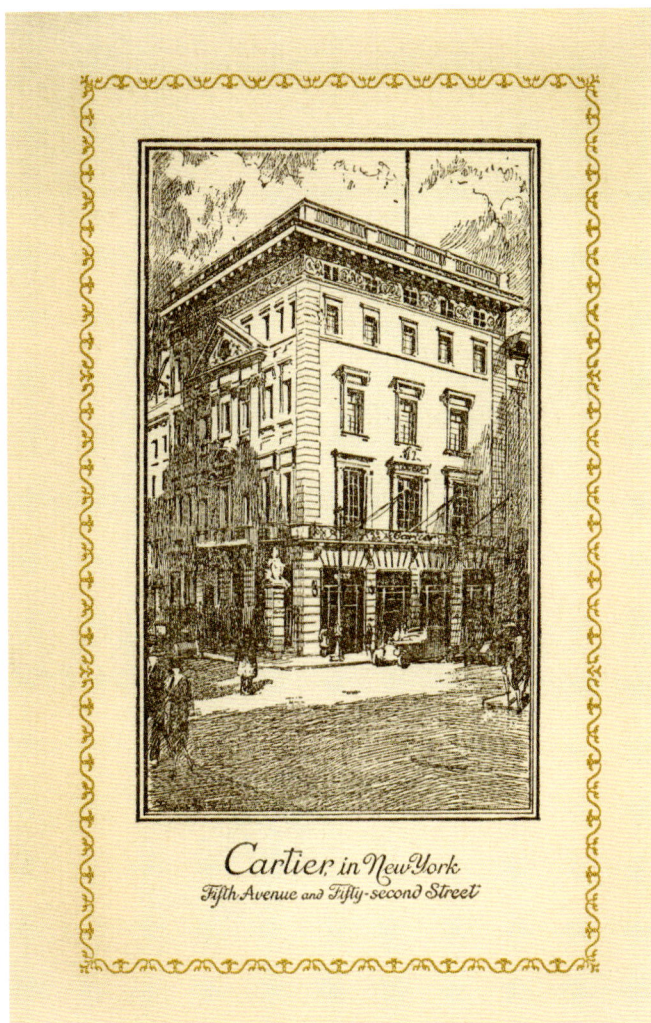

Cartier in New York
Fifth Avenue and Fifty-second Street

伦敦和纽约分部 (图 21) 迅速成为卡地亚国际战略的支点。巴黎、伦敦和纽约的三个分公司，分别由阿尔弗雷德·卡地亚的三个儿子执掌：整个家族都投入到了以卡地亚为名的国际拓展之中。

在保持整体密切协作的同时，每一个分店又都相互独立运营。每家分店都有自己的设计师和工作坊，以满足各地客户之间微妙的品位差异。

尽管如此，对日渐国际化的卡地亚客户而言，巴黎仍然具有无与伦比的重要意义。和平街成为法国奢侈品闪耀全世界的明星，并前所未有地成为珠宝与时尚之间的象征纽带。在皇室成员们借出访巴黎之机拜访卡地亚的时候，公众们也会蜂拥来到黑色的大理石门外，等候一睹国王与王后的华贵风采 (图 22、图 23)。

图 21 1927 年卡地亚位于纽约的精品店
Cartier Archives © Cartier

图 22 西班牙国王阿方索八世在 1920 年光临卡地亚巴黎精品店的盛况
Cartier Archives © Cartier

图 23 1925 年和平街 13 号卡地亚精品店的橱窗展示
Cartier Archives © Cartier

古老秩序与现代风格

　　古老的秩序在 1914 年轰然倒塌，以 18 世纪的法国为楷模的生活风尚也摇摇欲坠。在脱离了严格的社会等级和礼节之后，过去的庆典和仪式也荣光不再。第一次世界大战为这个处于优雅幻影的世界蒙上了一层黯淡的面纱 (图24、图25)。

　　工业革命令"美好年代"的生活态度和心智模式发生改变。服装也随之变化，以满足女性工作和运动的实用性需要。的确，这个时代的女性开始向往更广阔的自由，其中也包括不再约束其行动力的服装。保罗·波烈离开杜塞，转投沃斯麾下，设计出可在任何场合都穿着的简约服装。在他的影响之下，紧身衣消失了。随着日装和晚装的出现，花环风格也渐失风采。华丽的胸饰逐渐为各种以垂直线条为主的珠宝所取代。正是这些全新的形式，催生了 20 世纪20 年代的椭圆吊坠 (图26)。

图 24　罗马尼亚玛丽皇后佩戴卡地亚为其量身订制的冕冠
Cartier Archives © Cartier

图 25　玛丽皇后于 1928 年 5 月 15 日授予卡地亚的皇家委任状
Cartier Archives © Cartier

图 26 奥加尔·佩利公主
俄国沙皇亚历山大二世的第六个儿子
——大公保罗的妻子。她佩戴着由钻石和
珍珠装饰的帽饰和作为胸饰的镶有梨形
钻石的冠冕，于 1911 年制作。
Photo Boissonnas et Taponnier

　　这是一个高速发展的世纪，交通革命的年代，正如未来派作家
马里内蒂所预言，这个年代将拥有全新的生活准则。艺术也爆发出
缤纷的色彩。野兽派、维也纳分离派、德国表现主义、俄国芭蕾舞
纷纷登上舞台，令巴黎人为之惊叹与迷恋（图27）。

　　立体主义和抽象主义开始出现。路易·卡地亚在作品中引入了

明亮的色彩，以及富有创新、别具一格的几何造型。从简单的立方体，到多边形和菱形，均以彩色琢面宝石雕刻而成。

在此期间，卡地亚的另一大标志——腕表的制作也崭露头角。最早出现的是方形的 Santos 腕表。这一设计创新还带来了功能上的突破。腕表第一次兼顾到佩戴的舒适性与实用性，充分展示出设计如何可以在实现基本功能的同时仍不失创造性 (图 28)。

尽管潮流更迭变迁，在第一次世界大战结束之前，花环风格的冠饰却一直独享尊荣。对于无论是因为血统还是财富而跻身上流社会的精英人士而言，冠饰仍然是权力的象征。人们通常在典礼和社交场合佩戴冠饰，以彰显其尊贵地位。在 1911 年 4 月的第一个星期，数千名好奇的游客涌入了新庞德街，争相观赏陈列于卡地亚店中的 19 顶卡地亚冠饰。这些冠饰在两个月以后将在乔治五世和玛丽王后的加冕礼上被佩戴。

装饰艺术

随着中东欧皇室的没落，以及在战后重建的紧张经济局势之下，作为统治阶级之象征的冠饰也逐渐式微。冠饰在伦敦的衰落晚于巴黎。在那里，星期一和星期二的歌剧演出成为了盛大的珠宝展示庆典。璀璨的华光，为这些夜晚赢得了"王冠之夜"的雅号，尽管其中不无戏谑之意。

早在一战前夕，经典的冠饰就已经为东方风格的冠羽（图29）所取代，随后更是演变为简单的发饰。发饰是时尚淑女及服装设计师帕奎因和波烈最钟爱的配饰。20世纪20年代，发饰出现了丰富的几何造型和色彩组合，这正是典型的装饰艺术风格。在首饰变形风

图 29 铂金羽饰
羽毛状的铂金环凸显中央梨形的钻石（重19.42克拉）的夺目光辉，"羽毛"两边各有一枚较小的梨形钻石（每个重4.55克拉），1912年戈特查可夫王子委任卡地亚订制。
Cartier Archives © Cartier

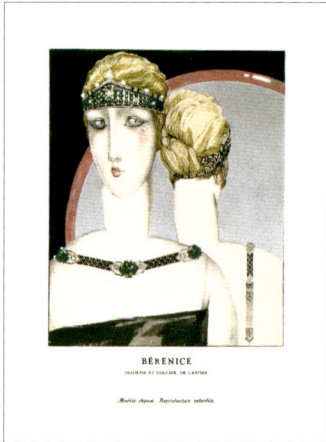

图 30　水彩画
这幅水彩画描述了一位佩戴伯妮斯发带和肩饰项链的模特，1925 年佩戴在身上的全套首饰在巴黎举办的装饰艺术和现代工业展上展出。
Cartier Archives © Cartier

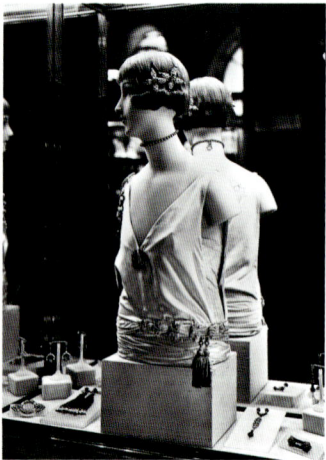

图 31　蜡塑模特
蜡塑模特所展示的是头饰（黑玛瑙和钻石制成的兰花，1923 年）以及用无色水晶、钻石和黑玛瑙制成的腰带。
Cartier Archives © Cartier

潮的影响下，这些发饰还可以被改为手镯和胸针。

新的风尚钟爱垂直线条。花环风格的飘逸弧线，为装饰艺术动感的硬朗线条所取代。来自古代东方文明的大胆造型和鲜活色彩大行其道。早在 1913 年，卡地亚就在自己的胸针、手镯和项链作品中，将镂空图案、伊斯兰星辰和菱形造型结合在了一起。

某些风格元素取材于古埃及的丧葬面具和植物图案。无独有偶，波斯风格的细密画和古董残片也以符合现代人审美的方式镶嵌其间。

卡地亚以自己独有的诠释方式，诠释了当时社会对失落文明的痴迷，以及对古代美学的回归（图 30、图 31）。

来自印度的浸润也催生了前所未有的珠宝设计，印度王公的传统珠宝和名贵宝石纷纷被拆解、重镶，创作成为全新的作品。印度风潮还通过压花丝绸、宫廷装饰和拉贾斯坦花园风格渗透到了珠宝领域。

覆盖前胸的多串式印度王公项链，也为卡地亚赋予了灵感，创作出彩色宝石雕刻项饰（图 32、图 33、图 34），以及 20 世纪 20 年代风行一时的悬垂至髋部的珍珠串饰。

这种浸润是相互的。一方面来自于西方人登上大吉岭特快列车探索印度时所感受到的异国情调，另一方面则来自于王公们对欧洲文化的浓厚兴趣。对印度古代珠宝或旅行纪念品加以改造，创作出集合了现代镶嵌手法和印度传统造型的独特作品。

黛丝·法罗斯、米希亚·塞特、埃尔茜·德·沃尔夫，乃至温莎公爵夫人等时尚标志及咖啡馆社交界的标杆人物均推动了印度首饰的流行（图 35）。在美国室内设计师埃尔茜·德·沃尔夫于凡尔赛家中举行的派对上，宾客们甚至见到了一头真正的来自斋浦尔的大象。

在卡地亚，贞·杜桑也是印度风格作品的引领者。她对印度风格的首饰有着天生的敏锐才华。自 20 世纪初起，她就一直佩戴这种风格的饰品。1933 年，杜桑重新启用黄金，打破了铂金对卡地亚珠宝风格和镶嵌工艺长达 30 年的统治。

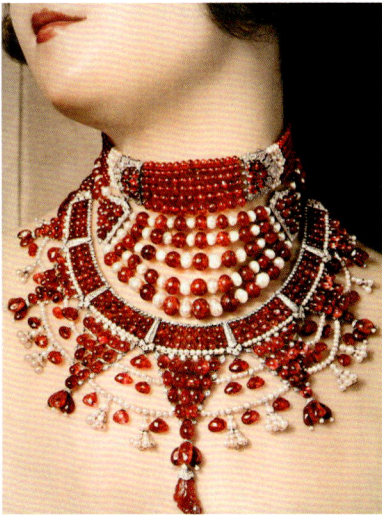

图 32　红宝石项链
红宝石项链是 1928 年卡地亚为印度土邦王公布品德拉·塞恩勋爵专门订制的。彩色胶片来自卡地亚图片典藏。红宝石、珍珠和钻石的项链是 1930 年为一位王公订制的。
Cartier Archives © Cartier

图 33　巴提亚拉巨型项链
1928 年为布品德拉·塞恩勋爵，即巴提亚拉王公订制。根据重量将宝石放在一个蜡制的盘子，戴比尔斯钻石(重 234.69 克拉)镶嵌在链坠的中心位置。
Cartier Archives © Cartier

图 34　佩带着巴提亚拉巨型项链的印度土邦王公雅达维德拉·塞恩
项链为其父布品德拉·塞恩勋爵于 1928 年订制，镶有戴比尔斯钻石。照片摄于 1940 年。
Cartier Archives © Cartier

全新方向

似乎是为了响应 1929 年的华尔街金融危机，一种全新的美学风潮也应运而生，宣告了飨宴般"咆哮的二十年代"的结束。珠宝首饰回归到了更加经典的风格，以不同切割方式的钻石组合出各种几何造型。这是一种通体白色的珠宝，由铂金、钻石和无色水晶组成，与 1920 年的富丽色彩形成了鲜明的对比。这是一个无论在政治上还是经济上均纷扰杂乱的时代，卡地亚的设计师和工艺师也在作品上做出了相应的调整。与此同时，他们也并没有抛弃卡地亚客户所一贯追求的柔美与魅惑的梦想。

图 35　晚宴
赛西尔·比顿和艾历克西斯德赫迪以及黛丝·法罗斯在艾克西斯·德赫迪男爵家中，即兰伯特酒店。
C. Beaton © Sotheby's London

图 36 1930 年的贞·杜桑
Cartier Archives © Cartier

图 37 卡地亚广告
Cartier Archives © Cartier

继贞·杜桑 (图36) 于 1933 年被任命为高级珠宝部门总监之后，黄金又以华丽瞩目的形象卷土重来。金成为卡地亚最钟爱的金属，不只是用作基座，同时也整体使用，创作出充满活力、富有雕刻感的首饰。

这种黄金首饰既抽象，又富有象征性，几乎毫无拘束，玩味着各种体积量感、重复手法和风格形式。部分作品镶嵌着彩虹般缤纷的宝石，有些甚至是前所未有的组合，比如将超现实主义的底色与紫水晶和海蓝宝石相搭配 (图37)。

在"杜桑风格"的影响下，卡地亚的动物意象作品也在 20 世纪 30 年代大放异彩。在第二次世界大战结束后的数年间，卡地亚的"动物王国"更是在贞·杜桑的悉心关注下进一步壮大。战争震撼了整个社会。大自然成为逃离人类自相残杀的终极避难所。动物和植物的世界超然于一切。

图 38 铂金和钻石制成的"天堂鸟"胸针
Cartier Archives © Cartier

自然永远不会遭到背叛。最多只会在现实主义与象征主义之间留出一定的距离。找到这种完美的距离，就是通往纵横创意的密钥。在卡地亚，镶嵌宝石的鸟类展开了它们象征和平的羽翼 (图 38)。

女性向往着更广阔的自由，立志寻求表现女性气质的全新形式。1947 年是充满新气象的一年，在女性的服装中，出现了纤细的腰线，柔和的肩部，以及巧妙包裹臀部的中长裙。精致就是一切。妆容更为清淡，也更为精致。这些女性呈现出了一种野性动物的灵息，仿如猫科动物的姿态。

卡地亚的第一枚立体豹形胸针出现于 1948 年，堪称一个名副

其实的动物雕塑。温莎公爵购得了这枚胸针，以及一对配套的耳环，赠送给温莎公爵夫人 (图39)。后者于第二年购入了自己的第二件豹形珠宝。这头豹蜷伏在一枚重达 152.35 克拉的凸圆形蓝宝石上，豹身镶满钻石，点缀刻面蓝宝石。眼神凌厉，令人着迷。

就在温莎公爵夫人为这个立体造型的现代首饰着迷之际，在伦敦冠饰依然是主要基调。在圣詹姆斯宫廷内，佩戴冠饰仍然是皇家礼仪的一部分。1953 年，伊丽莎白王太后还从卡地亚订制了一顶王冠 (图40)。尽管很多贵族名媛认为冠饰烦琐而过时，这一严格礼仪却丝毫没有动摇。

卡地亚则继续呼吸着时代的空气，基于反映风尚变迁的愿望和决心，创造出一种不断沿革的风范。卡地亚延续了其创始者的精神，以自己的作品去适应重大的经济异动和社会变迁。

色彩也是灵感之源。卡地亚创造了一种令人眩目的彩色宝石组合，将一种植根于历史的古老传统变成了永恒。不仅如此，卡地亚还在创造力的激发之下，创作出带有微妙差异的新作品，探索全新的色彩以及更加奇妙的对比。卡地亚还为其作品目录引入了全新的生物，以超现实主义和现代手法雕琢自己的神异世界，自然而然地阐释着一位卓越女性与装饰艺术的深情交融。

鲜活的记忆——卡地亚艺术典藏系列

卡地亚并不只是一间单纯的商业公司，而更是近似于一个铸造时代面貌的历史和文化机构。卡地亚的历史传承是其品牌个性的一个独特侧面。作为其事业的一部分，卡地亚收集、分类并保存着大量重要的文档。这是一个异彩纷呈的无价宝库，为珠宝创作赋予了全新的视角。因此，卡地亚将保存这些档案视为对文化机构、专业人士和艺术市场的义务与责任。

基于这一认识，卡地亚于 1983 年创建了卡地亚艺术典藏室。没

图 41 卡地亚档案资料
P Parinejad © Cartier

有任何东西能够取代"鲜活"的作品。它们是卡地亚的创造力、风格和专业工艺的终极体现。因此，长久以来，卡地亚一直致力于收集自家的创作珍品。并以风格和灵感，材质和技术对这些珍品加以甄选。

卡地亚对这些藏品具有深厚的感情。而这些藏品也正展示出了卡地亚对装饰装潢艺术的巨大贡献。

卡地亚保留了大量完美的原始文档，包括销售记录、创意文本、石膏模型、原始玻璃底片、研究、初步草图以及艺术家的终稿图纸（图41）。卡地亚艺术典藏室现在共搜罗了 1 360 件从卡地亚创立至今的珍贵作品。包括珠宝、冠饰、名贵的古董表、以及指针在无色水晶表盘内旋转，但却看不到内部隐藏机械的魅幻时钟。这个系列还包括一些器物和配饰，或铺张奢华，或简单实用，如化妆盒、装饰盒、书写工具和香烟盒等。

卡地亚这一行为迅速受到了众多顶级文化机构的欢迎。他们对于展示这样一个致力于保存一种风格，传承其艺术与特殊技艺的企业表现出了浓厚的兴趣，甚至可以说是极大的热忱。随着时间的推移，卡地亚艺术典藏系列展览的重心也在发生改变。最初是带有教育性质的回顾展，比如 1989 年在巴黎小皇宫举行的展览（图42）。1999 年在墨西哥城美术馆举行的展览则完全抛弃了教学和纪年目的，而是采取了更加自由的视角，根据灵感、色彩或材质来呈现艺术典藏系

列中的珍品。2002 年，在柏林维特拉设计博物馆举行的"索特萨斯眼中的卡地亚设计"则采用了一种全新的解读方式，将设计元素完全从其历史背景中抽离出来，第一次通过艺术家的视角来诠释一个珠宝品牌的作品 (图 43)。2009 年，在东京国立博物馆，日本设计师吉冈德仁从图像学角度对"卡地亚的创作记忆"加以了阐释。他认为"与美丽相逢的记忆"展览的视角可以被看作是一个实验性的场景，通过卡地亚作品及其背后的故事透视卡地亚的发展轨迹。

这些藏品由卡地亚从私人或拍卖会上购得，曾属于社会名流或普通人士。每一件都拥有其独特的历史。卡地亚艺术典藏系列是不同的历史年代及当代的见证，是珠宝历史和社会风潮的鲜活记忆与生动再现。

卡地亚形象、风格及传承总监

皮埃尔·雷诺

图 42 1989 ～ 1990 年在巴黎小皇宫举办大型回顾展"卡地亚艺术典藏系列"
James Lignier © Cartier

图 43 2002 年在柏林维特拉设计博物馆举行的"索特萨斯眼中的卡地亚设计"展览
E.Sottsass © Sottsass Associati / Cartier

JEWELLER OF KINGS, KING OF JEWELLERS

A pioneer in the history of jewellery, Cartier stands out as having paved the way for future trends, mastering new techniques and always following its intuition to develop a style that is truly its own.

   Throughout its history, this visionary jeweller has borne witness to how lifestyles have evolved, expressing these remarkable changes in its jewellery. History, the economy and society never stand still, and it is their constant evolution that inspires and nurtures Cartier's creativity.

fig.1 1900 Louis-François Cartier
In 1847 Louis-François Cartier (1819-1904) takes over the workshop of his master craftsman, Adolphe Picard, located at 29 rue Montorgeuil in Paris.
Cartier Archives © Cartier

From Saint-Eustache to the imperial court

La Maison Cartier was established in 1847 by Louis-François Cartier (1819-1904) (fig.1) , when he took over the business of his apprenticeship master Adolphe Picard, at 29 Rue Montorgueuil in the Saint-Eustache district of Paris.

As was usual at that time, he sold products from a number of other jewellery workshops.

In 1852, Louis-François Cartier moved to new premises at 5 Rue Neuve-des-Petits-Champs, behind the fashionable Palais Royal district, in the shadow of the imposing palace of the Orléans. His first distinguished customer was the Comtesse de Nieuwerkerke, an important figure of the arts under Napoleon III. Fashion was still influenced by the bourgeois taste of the July Monarchy. Cameos were very much in vogue, and so the Countess placed her first order with Cartier for an antique cameo necklace.

Her husband, the Comte de Nieuwerkerke, was the emperor's Superintendent of Fine Arts, and it was through the Countess's good offices that Cartier received his first commission from Princess Mathilde (fig.2) in 1856.

Three years later, Empress Eugénie herself ordered a silver tea service.

The shop on Rue Neuve-des-Petits-Champs had grown too small to accommodate a supplier to the imperial household. Hence Cartier moved again, this time to premises on Boulevard des Italiens (fig.3). On the eve of Haussmann's transformation of Paris, the boulevard was already a bustling thoroughfare. Ladies swept by in crinolines while crowds pressed into the cafés, ensuring a constant stream of customers. The cream of Parisian society under the Second Empire flocked to Cartier's.

Circumstances were less favourable when Alfred Cartier (fig.4) took over the business in 1874. The imperial family had sought refuge in England while the Commune had drained the city of its lifeblood.

Despite the uncertain economic and political climate, Cartier continued to sell a variety of articles, such as fans, Wedgwood medallions, jewellery, châtelaines and pocket watches. The company's stock registers (fig.5) show both a vast selection of jewellery and objets d'art in the prevailing Historicist style, alongside an array of traditional diamond jewellery.

Cartier's clientele grew. While the company's ledgers continued to record sales to France's noble families, they were now joined by growing ranks of bankers, industrialists and other affluent members of the business world.

The Cartiers wove special relations with the traditional aristocracy and well-to-do middle classes whose ostentatious spending betrayed their desire to be seen wearing the symbols of power.

fig.2 Princess Mathilde (1820-1904), niece of Napoléon I, introduced Cartier to Empress Eugénie in1859
© Collection Roger-Viollet

fig.3 Boulevard des Italiens
© Roger-Viollet

fig.4 Son of Louis-François Cartier, Alfred
Cartier (1841 - 1925) director from 1874.
Cartier Archives © Cartier

fig.5 record book
Cartier Archives © Cartier

This new and wealthy elite proved more international. When Cartier moved again, this time to the elegant Rue de la Paix, it was to an address in keeping with the expectations of this cosmopolitan clientele. Indeed, this prestigious artery was lined with the artisans of fashion: jewellers, couturiers, milliners, glovemakers, shoemakers and perfumers.

Rue de la Paix and the Garland style

The event that would facilitate the transition from a seller to a maker of jewellery came with the move to number 13 Rue de la Paix (fig.6), the heart of French fine jewellery. A design studio was set up in 1898, and in November 1899 Cartier's new shop was officially opened. This dual inauguration coincided with the emergence of a new type of jewellery, made from platinum in the Garland style.

Alfred made his eldest son Louis (1875-1942) a partner in the firm. A man of classical culture, and a collector of eighteenth-century furniture and objets d'exception, Louis Cartier (fig.7) would develop his own vision of jewellery, deliberately turning his back on the Art Nouveau style.

Cartier imposed platinum for its mounts. The metal was more solid, more luminous, but also harder to work with. The use of platinum ushered in a revolution in stone-setting techniques. This, along with new tools and ways of casting metal, produced jewellery with a natural, flowing drape. It also made it easier to sculpt lines and curves whereas silver, platinum's predecessor, had required heavy settings for stones. Another of silver's great disadvantages had been to dull diamonds' fire. Now that jewels were exposed to the recent invention of electric lighting, the illusion of sparkle could no longer be maintained as it had in the flickering glow of candlelight.

Impressions of lace, wrought ironwork and ornamental grilles created the transparent, insubstantial medium of the Garland style. Platinum would colour the forms jewellery would take, with bold interpretations of classical and neoclassical motifs, as captured in stylised floral designs, acanthus leaves,

fig.6 rue de la Paix store
Cartier Archives © Cartier

fig.7 Portrait of Louis Cartier (grand-son of the founder and Alfred's elder son) by Emile Friant, around 1905. Cartier Archives © Cartier

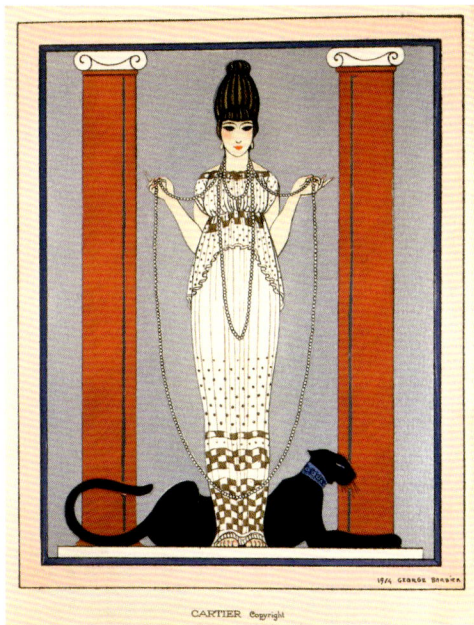

fig.8 watercolour by G. Barbier commissioned in 1914 by Louis Cartier for the invitation card to an exhibition organised at the salons in rue de la Paix. It was subsequently used as an advertisement in the 1920s and was published in the magazines of the period. First appearance of the panther. Cartier Archives © Cartier

garlands, wreaths, bows and droplets.

Following the influence of bourgeois taste, and the neo-Renaissance and neo-Baroque styles, the resurgence of the Louis XVI style benefited from a revival in classical tastes, as espoused by the traditional aristocracy but also the new social elite. Elaborate interiors were the setting for extravagant balls, lending a bygone air of ceremony to the gatherings of the aristocracy and the new moneyed class (fig.8).

fig.9 King Edward VII of England and
Queen Alexandra of England
Cartier Archives © Cartier

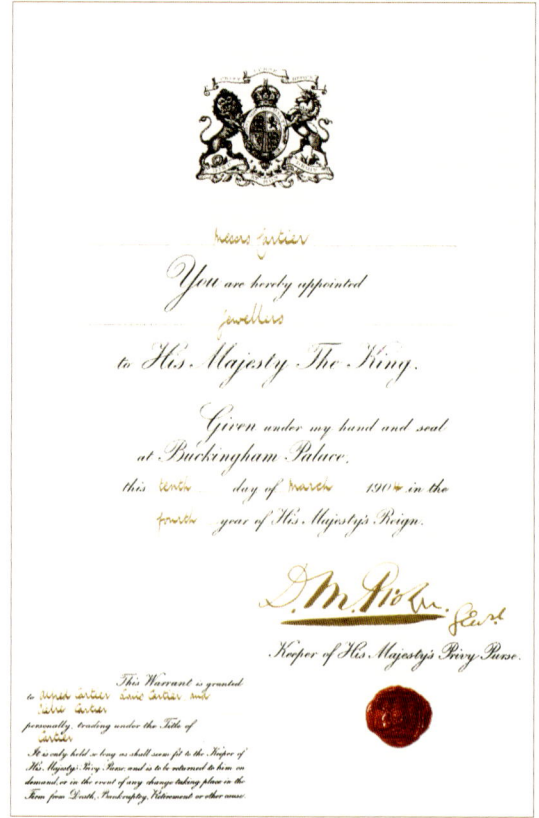

fig.10 10/03/1904 Warrant from King
Edward VII of England.
Cartier Archives © Cartier

fig.11 1904, Motif of the letterhead after
receiving the royal warrant as official
purveyor to King Edward VII of England.
Cartier Archives © Cartier

A new clientele, won over to classical tastes

Coming in the wake of industrial revolution, the Belle Époque was a period of unprecedented growth.

Means of transport and communication were modernised, the wealth of the colonies flooded into Europe, and captains of industry and bankers amassed huge fortunes, thanks as much to their entrepreneurial spirit as pure speculation.

Though lacking an aristocratic pedigree, their financial means were sufficient to give them access to the most closed circles. Now the trusted advisors of kings and emperors, they modelled their lifestyle and behaviour on the aristocracy.

Cartier did good business with this new elite, though the jeweller never neglected its customers in the courts of Europe and more distant realms.

Between 1904 and 1939, Cartier received fifteen letters patent, appointing it as official purveyor to different royal households. The first royal warrant was granted by King Edward VII of Great Britain in 1904 (fig.9, fig.10, fig.11). Others followed, including from King Alfonso XIII of Spain, Tsar Nicholas II, King George I of Greece, Queen Marie of Romania and the King of Siam.

To own a Cartier jewel guaranteed entry into a society in which maintaining one's rank depended as much on emulation and imitation as a constant quest for distinction. Hence Cartier supplied royal brides with their *corbeilles de mariage*. The extravagant luxury lavished on Marie Bonaparte's (fig.12, fig.13) *corbeille* reflected a taste for the symbols of an elite society; one that was equally evident in the trousseaux delivered to the bride by the grand *Maisons*. This illustrious descendant of Lucien Bonaparte epitomised the Empire style, as could be seen in the corsage ornaments, epaulettes and olive-leaf tiara which Cartier made for her in the spirit of her Napoleonic ancestor.

These creations were also typical of the Garland style that would establish the jeweller and ensure the international renown that would accompany its development.

fig.12 Marie Bonaparte who married the 2nd son of George I of Greece.
She wears on this photo her olive leaf diadem, in diamonds and platinum, commissioned for her wedding in 1907 with Prince George of Greece and Denmark.
Cartier Archives © Cartier

fig.13 Showcase of Cartier Paris for her wedding. Princess Bonaparte granted Cartier to display her numerous items in the showcases of the Rue de la Paix store in 1907.
Cartier Archives © Cartier

fig.14 Alfred Cartier and his three sons.
From left to right: Pierre, at the head of Cartier New York, Louis, at the head of Paris and Jacques, at the head of London.
Saint Jean de Luz, 1922.
Cartier Archives © Cartier

fig.15 1909 Advertising print presenting the three historical addresses Paris, London, New York.
Cartier Archives © Cartier

fig.16 Queen Alexandra of Great Britain wears a latticework diamond necklace, commissioned by Buckingham Palace from Pierre Cartier in 1904.
Portrait by François Flameng.
© The Royal Collection, Windsor

The three Cartier brothers and international expansion (fig.14, fig.15)

fig.17 Jacques Cartier, director of the London branch
Cartier Archives © Cartier

fig.18 1929 London Store
Cartier Archives © Cartier

On January 22nd, 1901, Queen Victoria died after a reign of almost sixty-four years.

After the mourning period at court had ended, England's leading families prepared for the coronation of Edward VII in an atmosphere of excitement. England was still a conservative country, and the jewellery proposed by Cartier's London branch created such a stir that the jeweller's workshops were overwhelmed with orders for tiaras, necklaces and corsage ornaments. Demand was such that Alfred Cartier sent his second son, Pierre (1878-1965), and his representative Alfred Buisson to ensure these commissions from England's nobility (fig.16) were honoured.

Cartier opened its London branch in 1902. Four years later, the youngest of Alfred Cartier's three sons, Jacques (1884-1942) (fig.17), took the reins of the London shop. He would remain at the head of 175-6 New Bond Street (Cartier's address as of 1909, after moving from 4 New Burlington Street) (fig.18) until the end of his days.

The London branch had barely opened than America beckoned. New York was home to more than three hundred millionaires. These were newly-

fig.19 Mrs Cornelius Vanderbilt with her diamond necklace
Cartier Archives © Cartier

fig.20 Pierre Cartier, director of the New York branch
Cartier Archives © Cartier

made fortunes. Granted, families such as the Morgans, the Vanderbilts and the Leeds rested on well-established financial assets; their wealth nonetheless enrolled them into a new social hierarchy.

From the sumptuous mansions along Fifth Avenue to the boxes of the Metropolitan Opera House, ambitious society hostesses vied for privileges and distinctions, and ostentatious spending, in particular on precious stones, was a favourite weapon in their campaigns (fig.19).

In 1909, Cartier opened its American branch, intent on winning the loyalty of the New World's social and economic elites, who were already long-standing patrons of the Parisian jewellery house. Pierre Cartier (fig.20) was entrusted with the future of the shop on 712 Fifth Avenue. In 1917, Cartier transferred its New York business to a Renaissance-style mansion at 653 Fifth Avenue, which the jeweller bought from Morton Plant. The transaction was nothing if not unusual, as the house was symbolically exchanged for a two-strand necklace of fifty-five and seventy-three fine pearls which had caught the eye of Morton Plant's young wife.

Cartier's London and New York branches (fig.21) rapidly became anchor points in an international strategy. The three subsidiaries of Paris, London and New York were managed by Alfred Cartier's three sons: an entire family

Cartier in New York
Fifth Avenue and Fifty-second Street

LONDRES
175-176, New Bond St. W.

NEW-YORK
653, Fifth Avenue V. S.

Cartier
13, Rue de la Paix
Paris

devoted to the international expansion of the jeweller that bore its name.

While forming an extremely tight-knit whole, each branch was run independently of the others. Each had its own designers and workshops which catered to the subtle differences in tastes of their particular cosmopolitan clientele.

Paris nonetheless remained of paramount importance for the customers of a jeweller which had grown into a worldwide institution. Rue de la Paix was the shining star of French luxury throughout the world and, more than ever, symbolised the link between jewellery and fashion. Should crowned heads on an official visit to Paris come to Cartier, crowds would throng to catch a glimpse of the kings and queens as they alighted outside the shop's black marble frontage (fig.22, fig.23).

fig.21 The Cartier New York store in 1927
Cartier Archives © Cartier

fig.22 King Alfonso XIII of Spain visiting Cartier Paris in 1920.
Cartier Archives © Cartier

fig.23 Window shopping 13 Rue de la Paix. 1925
Cartier Archives © Cartier

The old order and Modern style

A lifestyle modelled on eighteenth-century France held sway until the old order began to crumble in 1914. Bereft of the strict social hierarchy and etiquette on which they relied, the festivities and ceremonial inherited from the past were no longer relevant. The First World War cast a dark veil over this world of elegant illusion (fig.24, fig.25).

Industrial revolution caused attitudes and mentalities to change in the Belle Époque. Clothes began to adapt to the practicalities of work and women's sport. Indeed, women aspired to greater freedom, and this included fashions that no longer restricted their movements. When Paul Poiret left the couturier Doucet to join Worth, he designed simple day dresses to wear on the omnibus or in the Métropolitain. It was through his influence that the bodice disappeared. As new fashions for day and evening wear appeared, little by little the Garland style fell out of favour. Imposing corsage ornaments were gradually replaced by a different type of jewellery, dominated by vertical lines. These new forms would pave the way for the oblong pendants of the 1920s (fig.26).

fig.24 Queen Maria of Romania with one of her tiara from Cartier
Cartier Archives © Cartier

fig.25 15/05/1928 warrant from Queen Maria of Romania
Cartier Archives © Cartier

This was the century of speed, the revolution of transport and, as the Futurist writer Marinetti predicted, it lived by new rules. Art was an explosion of colour, with the Fauves, the Vienna Secession, German Expressionism and of course the Ballets Russes, which left Parisians both fascinated and astounded (fig.27).

Cubism and Abstraction appeared. Louis Cartier introduced bright colours and innovative, stylised, geometric forms, from simple cubes to polygons and lozenges sculpted from calibrated coloured stones.

This was also when another aspect of Cartier's identity – form watches - would assert itself, beginning with the square shape of the Santos watch. This innovation in design was also groundbreaking in terms of function. For the first time ever, comfort and practicality prevailed in a watch designed to be worn on the wrist. Cartier showed how design could achieve the essential and also be creative (fig.28).

Despite such revolutions, Garland-style tiaras enjoyed special status until the end of the First World War. Reserved for an elite, whether by birth or by fortune, the tiara remained a symbol of power, worn at ceremonies and social gatherings to denote a privileged status.

In the first week of April 1911, thousands of curious visitors flocked to New Bond Street to admire the nineteen Cartier tiaras which, two months later, would be worn at the coronation of King George V and Queen Mary.

Art Deco

As central and eastern European empires lived their last days, and amidst the strained economic climate of post-war reconstruction, the tiara, a symbol of the ruling classes, disappeared. This decline came later to London than Paris, where Monday and Tuesday evenings at the opera had become extraordinary processions of jewellery. Such glittering parades had earned these evenings the name, not without irony, of *soirées de diadèmes* or tiara nights.

Even before the outbreak of the Great War, the classic tiara had been supplanted by aigrettes in an Oriental style (fig.29), and later by a simple bandeau. They ranked among the accessories favoured by fashionable ladies and their couturiers, Paquin and Poiret. The 1920s bandeau bore witness to the rich potential of geometric forms, and to the influence and colour

fig.29 Platinum aigrette, featuring a navette diamond, a pear-shaped diamond (19.42 carats) and two other pear-shaped diamonds (4.55 carats each), commissioned by Prince Gortchakoff in 1912.
Cartier Archives © Cartier

combinations that would define the Art Deco style. Embracing the fashion for transformable jewellery, these bandeaux could instantly become bracelets and brooches.

New fashion favoured vertical lines. The suspended curves of the Garland style gave way to the plummeting lines of the Art Deco movement. Daring forms and vibrant colours ruled the day, borrowed from ancient Eastern civilisations. As early as 1913, Cartier incorporated openwork motifs, Islamic stars and lozenges into its brooches, bracelets and necklaces.

Certain stylised elements could be likened to Ancient Egyptian funeral masks and plant motifs. Similarly, Persian miniatures and antique fragments were set in mounts that were pleasing to the modern eye.

These were contemporary expressions of a fascination with lost civilisations, and a return to ancient aesthetics which Cartier interpreted in its own way (fig.30, fig.31).

The influence of India produced unprecedented jewellery designs, as the traditional jewels and precious stones of the maharajahs were dismantled and remounted as new creations. India's influence also crept into jewellery through patterned silks, palatial ornamentation and the gardens of Rajasthan.

The maharajahs' necklaces, whose multiple strands covered the entire chest, inspired the sautoirs of carved coloured stones and the strands of pearls that hung down to the hips that would dominate 1920s fashion (fig.32, fig.33, fig.34).

This was a mutual influence, shared between the exoticism that came with the discovery of India aboard the Darjeeling Express, and the interest which the maharajahs themselves nurtured for European culture. Antique jewellery or souvenirs of a voyage to India were transformed into unique creations, marrying contemporary settings with traditional Indian forms.

Such fashion icons and eminent members of café society as Daisy Fellowes, Misia Sert, Elsie de Wolfe, and even the Duchess of Windsor spread the fashion for Indian jewels. At a party hosted by the American interior designer Elsie de Wolfe, at her home in Versailles, guests were treated to the sight of real elephants from Jaipur (fig.35).

At Cartier, Jeanne Toussaint was at the head of the jeweller's Indian line. She had a natural flair for jewellery in the Indian style, for the simple reason she had been wearing it since the 1910s. In 1933, she ended platinum's thirty-year reign over the style and setting of Cartier jewellery when she reinstated yellow gold.

fig.32 Ruby necklace made to order for Maharaja Bhupinder Singh of Patiala by Cartier, Paris in 1928. Autochrome plate from the Cartier photographic collection.
Ruby, pearl and diamond necklaces created for a Maharani in 1930.
Cartier Archives © Cartier

fig.33 Study for the setting of the stones of a necklace for Sir Bhupindar Singh, Maharajah of Patiala created in 1928. The stones are put on a wax plate and grouped according to their weight. The "de beers" diamond of 234.69 carats is the centerpiece of the pendant.
Cartier Archives © Cartier

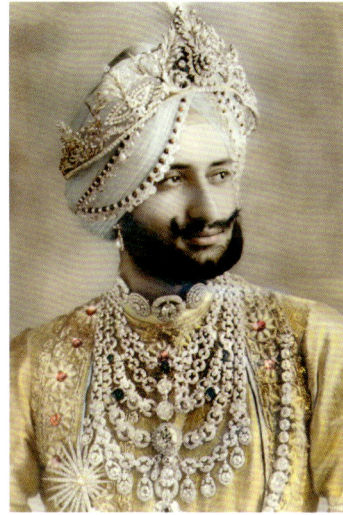

fig.34 The Maharajah Yadavindra Singh of Patiala, wearing the necklace, created for his father sir Bhupindra Singh in 1928. circa 1940. His Royal Highness the Maharajah Yadavindra Singh of Patiala, wearing the necklace set with the "de beers" diamond, created by Cartier for his father, Sir Buhupindra Singh, 1928.
Cartier Archives © Cartier

A new direction

As though echoing the 1929 Wall Street Crash, a new aesthetic marked the end of the euphoric Roaring Twenties. Jewellery returned to a more classic style, building on combinations of diamonds with different cuts to create geometric forms. Contrasting with the exuberant colours of the 1920s, this was all-white jewellery made with platinum, diamonds and rock crystal. These were troubled times, both politically and economically, and Cartier's designers and artisans adapted their work accordingly, though never forgetting the dreams of glamour and femininity that the jeweller's customers sought.

Following Jeanne Toussaint's (fig.36) appointment, in 1933, at the head of the fine jewellery department, yellow gold returned as imposing, massive jewellery. Gold was now the jeweller's preferred metal, no longer used solely in settings but to express its full measure in vibrant,

fig.35 Cecil Beaton, Alexis de Rédé and Daisy Fellowes during a dinner at the Hotel Lambert, home of Baron Alexis de Rédé. Courtesy Thierry Coudert.
C. Beaton © Sotheby's London

fig.36 Jeanne Toussaint circa 1930
Cartier Archives © Cartier

fig.37 Cartier advertisement
Cartier Archives © Cartier

sculptural jewels.

Part abstract, part figurative, impossible to pin down, this yellow gold jewellery played on volumes, repetitions and stylised forms. Certain pieces were set with fine stones in rainbow colours, some in unheard-of combinations such as amethysts and aquamarines with surrealist undertones (fig.37).

Cartier's bestiary gained prominence in the 1930s, again under the influence of *le style Toussaint*. In the years that followed the Second World War, under Jeanne Toussaint's careful eye, the jeweller's animal kingdom grew.

War shook society to its core. Nature appeared as the ultimate refuge from man's brutality to man; the animal and plant realms took over.

Nature would not let herself be betrayed. At most, a certain distance was allowed, between realism and symbolism. Finding this perfect distance was the key to unfettered creativity. At Cartier, jewelled birds spread their wings as symbols of peace (fig.38).

Women aspired to greater freedom and set out to conquer a new form of femininity. In 1947, the year of the New Look, women's silhouettes sported

fig.38 Brooch 'Bird of Paradise' in
platinum and diamonds
Cartier Archives © Cartier

a wasp waist, soft shoulders and longer skirts that cleverly skimmed the hips. Sophistication was the key. Make-up was less heavy, more subtle. These women had animality, the feline's poise.

Cartier's first three-dimensional panther brooch, a veritable animal sculpture, appeared in 1948. The Duke of Windsor acquired it, with a pair of matching earrings, for the Duchess (fig.39), who the following year took possession of her second panther jewel. Crouched on a cabochon sapphire of 152.35 carats, paved with diamonds and flecked with calibrated sapphires, its ferocious gaze holds us enthralled.

While the Duchess of Windsor gave in to the temptation of the daring three-dimensional forms of modern jewellery, in London the tiara still set the tone. At the Court of St James's, protocol still required that tiaras be worn. As

late as 1953, Queen Elizabeth the Queen Mother (fig.40) received a tiara from
Cartier. Although regarded by many aristocratic ladies as a nuisance and out
of keeping with the times, the strict rules of etiquette were unwavering.

And so Cartier has continued to live and breathe its era, and from this
desire and determination to reflect changing lifestyles comes a style in
constant evolution. True to its founders' spirit, through its creations, Cartier
apprehends the great economic and social transformations.

Colours bring inspiration. Cartier innovates with dazzling combinations
of coloured stones, thus perpetuating a tradition that takes root in its past.
More than this, creativity inspires the jeweller to introduce new nuances, and
to search for new colours and even more stunning contrasts. The jeweller
also welcomes new creatures into its fold, sculpting its menagerie with
extraordinary realism and modernity to nurture the dialogue between a
woman and the art of embellishment.

fig.40 Queen Mother wearing a Cartier tiara
Cartier Archives © Cartier

The Cartier Collection, a living memory

Cartier cannot be considered purely and simply as a commercial venture. Its stature is akin to that of a historic, cultural institution, one that helps shape its times. Cartier's heritage is a distinct facet of its identity. As part of its vocation, Cartier has assembled an extraordinary collection of documents, which it classifies and conserves. Remarkable for its diversity, this inestimable treasure sheds new light on jewellery creation. For this reason, Cartier maintains this body of documents as a duty and an obligation towards cultural institutions, specialists and the art market.

Mindful of this duty, the Art of Cartier Collection was created in 1983. Because nothing can replace the object "in the flesh" - for it is the ultimate

fig.41 Cartier archives
P Parinejad © Cartier

demonstration of the creativity, style and expertise of Cartier – over time, the jeweller has gathered together examples of its work. They are chosen for their style and inspiration, and for the materials and techniques brought into play.

These pieces are of immense interest to Cartier, but also reveal the jeweller's vast contribution to the decorative arts as a whole.

Inventoried from an extensive collection of original archive documents, comprising sales registers, ideas books, plaster casts, original glass negatives, studies, preliminary sketches and artists' final renderings, the Cartier Collection now numbers over 1,360 pieces from the origins of Cartier to the present day (fig.41). It brings together jewellery, tiaras, precious and classic watches and mystery clocks whose hands turn inside a rock crystal dial with no visible link to the hidden mechanism. The collection also includes extravagant or simply practical objects and accessories such as vanity cases, decorated boxes, writing instruments and cigarette cases.

This initiative was immediately welcomed by the most prestigious cultural institutions. They were seduced, and in many cases enthused, by this opportunity to present a company so attached to perpetuating a style and to passing on its artistry and exceptional expertise.

Over time, exhibitions of the Collection have shifted in emphasis. The very first were retrospectives with an educational vocation, such as at the Petit Palais in Paris (fig.42), in 1989. In 1999, at the Museo de Bellas Artes in Mexico City, pedagogical and chronological considerations were swept aside, making way for a much freer vision as pieces from the Collection were presented according to their inspiration, colours or materials. In 2002, *Cartier Design viewed by Ettore Sottsass* at the Vitra Design Museum (fig.43) in Berlin drew

on a new way of seeing the pieces, as elements of design independently of their historical context. For the first time an artist gave his vision, his interpretation, of a jeweller's work. In 2009, at the Tokyo National Museum, the Japanese designer Tokujin Yoshioka gave an iconographic interpretation of the Memory of Cartier Creations. His scenography for the exhibition *Story of …* could be interpreted as an experimental vision of Cartier's trajectory, seen through its creations and the story each one tells.

Purchased by Cartier from private sellers or at auction, these objects belonged to famous people or ordinary individuals. Each has its own history. Bearing witness to different eras, including the most contemporary, the Cartier Collection is both a living memory and a vivid recreation of the history of jewellery, and that of a society in constant movement.

fig.42 Musee du Petit Palais during the Art of Cartier Exhibition 1989-1990 James Lignier © Cartier

fig.43 E. Sottsass © Sottsass Associati / Cartier

Image, Style and Heritage Director
Cartier

Pierre Rainero

图版目录

LIST OF PLATES

图 版

PLATES

服务欧洲皇室

Contribution to the European Royal Courts
and the High Society

1847 年，路易·弗朗索瓦·卡地亚（1819 ～ 1904 年）从他师傅手中，接下了位于巴黎蒙特吉尔街的珠宝工作坊。路易与儿子阿尔弗雷德（1841 ～ 1925 年）一起，将这个小小的工作坊的名字——卡地亚变成了一个国际知名的珠宝品牌，并成为众多皇室贵族的御用供应商。1899 年，卡地亚迁址至尊贵的巴黎和平街 13 号，与沃斯、杜赛、娇兰等诸多顶级品牌为邻。

　　卡地亚最初的作品仍然具有清教徒式的风格和路易·菲利普国王统治时期（1830 ～ 1848 年）典型的布尔乔亚特色。在令人眼花缭乱的法兰西第二帝国时期（1852 ～ 1870 年），卡地亚迅速成长，并很快获得了皇室的赏识。以古典风格或古董为灵感的珠宝作品（参见 JS 01 C1850），以及各种精美的物品、银器和名贵钟表，为越来越多优雅和富有的顾客所钟爱，其中不乏玛蒂尔德公主、乌韦克尔克伯爵夫妇、索尔提科夫王子、比贝斯哥王子以及巴黎伯爵等贵族名流。1859 年，路易·弗朗索瓦·卡地亚终于梦想成真：拿破仑三世的妻子欧珍妮皇后请卡地亚为国王订制了一套纯银茶具。1907 年，玛丽亚·费奥多萝芙娜皇后亲自拜访了位于和平街的卡地亚殿堂。同年，卡地亚还接到了沙皇尼古拉斯二世的皇家委任状。从 1904 到 1939 年间，卡地亚一共收到了 15 份为皇室和宫廷成员担任御用珠宝商的委任状。温莎公爵夫妇从他们婚姻伊始到生命结束一直是卡地亚的忠实顾客。

　　19 世纪 60 年代末期，南非大型钻石矿床的发现为珠宝交易开创了全新局面。银行家、工业巨子，以及来自美国、英国和德国的投机商人纷纷效仿皇室贵族，订购比肩皇室珠宝的饰品。

　　1898 年，阿尔弗雷德的长子路易（1875 ～ 1942 年）成为卡地亚的合作伙伴进入卡地亚。卡地亚在他的决策下于 1899 年搬迁到了和平街。路易是一个唯美主义者，具有超乎寻常的商业直觉和战略眼光。他崇尚 18 世纪的法国文化，尤其是绘画和极为优雅的装饰艺术。路易一生收藏了大量 18 世纪的器物和家具，他的审美品位极大地影响着卡地亚的未来创作，开创了镶嵌于铂金基座的全新珠宝风格，以璀璨闪亮的铂金取代沉重的金银基座（参见 HO 08 A02），创造出极富现代感的珠宝（参见 CL 122 A12, CL 114 A03, NE 39 A06）。铂金工艺为珠宝赋予了一种自然流畅的褶皱造型，更易于进行灵动的直线和曲线塑造，具有超乎寻常的强度和柔韧性，可以让设计师实现各种大胆的设计。"种子式"或"串珠式"镶嵌法就是将钻石精致地镶嵌于铂金基座之中，呈现出形似小珠的规则锯齿，为珠宝赋予了极致柔美的触感。1906 年的百合花三角胸衣胸针（参见 CL 134 A06）就是其中最精美华丽的代表作，卡地亚将这一独特技术发挥到极致，即使是纤细的花蕊上也装饰着种子式镶嵌的玫瑰式切割钻石。这种风格后来被称为"花环风格"，彰显了卡地亚的制作功力，不仅为卡地亚创造了与众不同的独特风格，更因此被英国国王爱德华七世赞誉为"皇帝的珠宝商，珠宝商的皇帝"。

卡地亚艺术典藏室馆长　帕斯卡尔·勒博

In 1847, Louis François Cartier (1819-1904) bought back the workshop of his mentor, located in the Rue Montorgueil in Paris. Along with his son, Alfred (1841-1925), he transformed a modest workshop into an internationally renowned jeweller, and official supplier to most of the Royal and Princely Courts. It was then, in 1899 that Cartier moved to the, now prestigious, address of 13 Rue de la Paix in Paris, where the jeweller was in distinguished company, from the couturier Worth to Doucet and Guerlain.

The initial works of the firm are still marked by the puritan spirit and bourgeoisie characteristics of the reign of King Louis Philippe (1830-1848). During the dazzling Paris of the Second Empire (1852-1870) Cartier grew, soon being recognized by the Imperial Court. Jewels of classical or antique inspiration (see JS 01 C1850) were placed alongside delicate objects, silverware, precious watches and clocks, being offered to an exceedingly elegant and wealthy clientele, including Princess Mathilde, the Count and Countess of Nieuwerkerke, Prince Soltikov, Prince Bibesco, and the Count of Paris. In 1859 one of Louis- François's dearest wishes came true, when the wife of Napoleon III, Empress Eugénie, ordered a silver tea service from him. In 1907 Empress Maria Feodorovna visited Cartier's establishment in the Rue de la Paix. That same year Cartier would receive the Imperial warrant of Czar Nicholas II of Russia. Between 1904 and 1939 fifteen patent letters appointed Cartier official purveyor to Royal and Princely Courts. From their marriage in 1937 to their death, the Duke and the Duchess of Windsor remained loyal clients to the Maison.

The discovery of major diamond deposits in South Africa in the late 1860s brought a new dimension to the jewellery trade. Bankers, industrialists, and speculators from America, England, and Germany joined the ranks of the aristocracy, ordering jewellery fit for an emperor.

In 1898, Alfred's eldest son, Louis (1875-1942), eventually became his business partner. Louis held a clear influence in the decision to move to the Rue de la Paix premises in 1899, and was simultaneously an aesthete and a shrewd strategist with good business instincts. He admired eighteenth-century French culture, notably painting and the highly elegant decorative arts. Throughout his life he amassed a collection of exceptional eighteenth-century objects and furniture, his taste influencing Cartier's future creativity. This influence coincided with a new style of jewels; replacing heavy gold and silver mounts (see HO 08 A02) with glittering threads of stainless metal, bringing forth a modern style of jewellery (see CL 122 A12, CL 114 A03, NE 39 A06). Platinum processing endowed a natural and smoothly draped form to gems, easing the difficulty of linear and curve production. The extreme strength and malleability of platinum allowed jewellers to create an assortment of daring designs: One of the traits of these jewels was the "millegrain" or "beaded" setting: delicately mounted in platinum with regular indentations that resembled small pearls giving a supremely refined touch to the diamond-set jewels. The lily stomacher brooch made in 1906 (see CL 134 A06) is perhaps the most magnificent example. Here Cartier took technical virtuosity to its extremes: for even the tiny stamens are adorned with millegrain-set rose-cut diamonds. The "Garland style," as it would later be called, through Cartier's design and technical virtuosity not only made it incomparable to other jewellry, but even inspired King Edward VII to pronounce Cartier " Jeweller of Kings, King of Jewellers".

Pascale Lepeu
Curator of the Cartier Collection

I

JS 01 C1850

全套首饰

约 1850 年
卡地亚巴黎
金
椭圆形和梨形切面紫水晶
玳瑁
项链　42.5 厘米
发梳　11.3 × 12.0 厘米
胸针　8.0 × 4.0 厘米
耳坠　4.0 × 1.7 厘米

I

JS 01 C1850

Parure

Cartier Paris, c.1850
Gold
Oval-and pear-shaped faceted amethysts
Tortoiseshell
Necklace 42.5 cm;
comb 11.3 × 12.0 cm;
brooch 8.0 × 4.0 cm;
pendant earrings 4.0 × 1.7 cm

2

CL 287 C1860

贝雕胸针

约 1860 年
卡地亚巴黎
錾刻及抛光金
贝壳浮雕
6.28 × 5.05 × 2.21 厘米

这枚胸针上面带有路易·弗朗索瓦·卡地亚的注册标志：一个两侧带首字母缩写"LC"的红桃A。

2

CL 287 C1860

Cameo brooch

Cartier Paris, c.1860
Chased and polished gold
Shell cameo
6.28 × 5.05 × 2.21 cm

This brooch bears the maker's mark registered by Louis François Cartier: an ace of hearts flanked by the initials LC.

3

OV 09 C1860

糖碗

约 1860 年
卡地亚巴黎
银镀金
刻花玻璃
13.0 × 12.2 厘米

3

OV 09 C1860

Sugar bowl

Cartier Paris, c. 1860
Silver-gilt
Engraved glass
13.0 × 12.2 cm

手镯

约 1865 年
卡地亚
金，银
青金石
玫瑰式切割钻石
3.75 × 1.33 厘米

委托订制这款手镯的客人表达了他对君主制度的拥护，"百合花"是法国君主的象征。

4
BT 125 C1865

Bracelet

Cartier, c. 1865
Gold, silver
Lapis lazuli
Rose-cut diamonds
3.75 × 1.33 cm

The client who commissioned this bracelet had displayed his monarchist opinions, the "fleur de lys" being the symbol of French monarchy.

"鲁昂十字架"吊坠

约 1865 年
卡地亚巴黎
玫瑰金
人造宝石
14.00 × 7.75 厘米

镂空金丝"鲁昂十字架"是一种传统首饰。人造钻石约兴起于1750年，通常被用于镶嵌宝石十字架。依照传统，订婚的青年男子通常会在本地市场"寻觅首饰"，购买项链、链绳和十字架。

5

PE 12 C1865

Croix de Rouen pendant

Cartier Paris, c.1865
Pink gold
Paste
14.00 × 7.75 cm

A "cross of Rouen" in openwork gold was a traditional item of jewellery. Paste diamonds, developed around 1750, were frequently used for crosses set with stones. Young men engaged to be married would traditionally "go jeweling" at regional fairs to buy necklaces, chains, and crosses.

6

JA 15 C1870

腰带扣

约 1870 年

卡地亚巴黎

银

8.50 × 4.30 × 1.00 厘米

6

JA 15 C1870

Belt buckle

Cartier Paris, c. 1870

Silver

8.50 × 4.30 × 1.00 cm

7

FK 20 A1873

香精瓶

1873 年
卡地亚巴黎
金，银
水晶
绿松石色珐琅
12.80 × 3.90 × 2.60 厘米

7

FK 20 A1873

Scent bottle

Cartier Paris, 1873
Gold, silver
Crystal
Turquoise-coloured enamel
12.80 × 3.90 × 2.60 cm

WB 24 A1874

珐琅腰链表

1874 年
卡地亚巴黎
黄金，錾刻玫瑰金和珐琅装饰（腰链）
田园牧歌场景彩色珐琅，蓝色、白色珐琅
珍珠
16.5 × 3.4 厘米

腰链是17世纪男性和女性佩戴于腰间的一种长链。这种腰链通常以装饰性的嵌片和各种链节组成。链上坠以风格匹配的钟表，通常还搭配多个坠饰。这种配饰极富意趣、富丽华贵。这款腰链完好保留在原盒中，是此类华丽精致饰品的完美代表。

此表为圆形机芯，镀金，圆柱形擒纵系统，单金属平衡摆轮，扁平摆轮游丝。铰链式镜面，可开启调校时间或上链（4点钟位置）。

8

WB 24 A1874

Enameled watch on chatelaine

Cartier Paris, 1874
Yellow gold, chased pink gold and
enameled decoration (chatelaine)
Polychrome enamel showing bucolic
scenes, blue and white enamel
Pearls
16.5 × 3.4 cm

A chatelaine is a long chain worn since the 17th century at the waist by both men and women. Such chains were composed of ornamental plaques joined by various kinds of link. Attached to the chain would be a matching watch, often accompanied by several charms. This inspiring piece of decorative finery could attain a high degree of magnificence. Surviving with its original presentation case, this chatelaine is a perfect example of these richly elaborated pieces.

Round movement, gold-plated, cylinder escapement, monometallic balance, flat balance spring. Hinged crystal opens to allow access for setting the time and winding the watch (at 4 o'clock).

9

WB 37 C1880

表胸针

约 1880 年

卡地亚

镂空雕花及抛光金

6.0 × 3.45 厘米

为可分离式三叶草结吊坠。

此表为圆形机芯，镀金，圆柱形擒
纵系统，扁平摆轮游丝。

9

WB 37 C1880

Watch-brooch

Cartier, c. 1880

Chased, openwork and polished gold

6.0 × 3.45 cm

The trefoil pendant bow is detachable.

Round movement, gold-plated,
cylinder escapement, flat balance
spring.

10

BC 09 C1880

袖扣（一对）

约 1880 年

卡地亚巴黎

玫瑰金，黄金

直径　2.40 厘米

描绘障碍赛马场景的灰色玛瑙浮
雕。

10

BC 09 C1880

Pair of cufflinks

Cartier Paris, c. 1880

Pink gold, yellow gold

Diameter 2.40 cm

Gray agate cameo depicting steeple-
chase scenes.

托盘式汤碗（一对）

约 1890 年
卡地亚
錾刻银
汤碗　9×20×12 厘米
碟　22.50×2 厘米

II

SI 04 C1890

Pair of bouillons (covered cups)

Cartier, c. 1890
Chased silver
Cups 9 × 20 × 12 cm; saucers 22.50 × 2 cm

12

CL 291 A1894

牵牛花胸针

1894 年
卡地亚巴黎
金，银
半透明红色、乳白色珐琅
玫瑰式切割钻石
2.8 × 2.58 厘米

12

CL 291 A1894

Convolvulus brooch

Cartier Paris, 1894
Gold, silver
Translucent red and opalin enamel
Rose-cut diamonds
2.8 × 2.58 cm

13

CL 270 A1898

蝴蝶结胸针

1898 年
卡地亚巴黎
金，银
旧式切割和玫瑰式切割钻石
3.50 × 3.40 厘米

13

CL 270 A1898

Ribbon bow brooch

Cartier Paris, 1898
Gold, silver
Old- and rose-cut diamonds
3.50 × 3.40 cm

I4

BT 17 C99

手链

约 1899 年
卡地亚巴黎
金，银
玫瑰式切割钻石
六颗枕形红宝石
长度　18.0 厘米

I4

BT 17 C99

Bracelet

Cartier Paris, c.1899
Gold, silver
Rose-cut diamonds
Six cushion-shaped rubies
Length 18.0 cm

15

PE 03 C1900

吊坠

约 1900 年
卡地亚巴黎
金，银
圆形旧式切割和玫瑰式切割钻石
方形、长方形、圆形切面蓝宝石
圆形切面、圆柱形切割红宝石
花式切割祖母绿
一颗天然珍珠
9.0 × 4.8 厘米

这款文艺复兴风格的吊坠来自列昂·库隆工作坊。该工作坊于1884年成为卡地亚的供应商。

15

PE 03 C1900

Pendant

Cartier Paris, c.1900
Gold, silver
Round old- and rose-cut diamonds
Square, rectangular and round faceted
sapphires
Round faceted and calibré-cut rubies
Fancy-cut emeralds
One natural pearl
9.0 × 4.8 cm

This Renaissance-style pendant comes from the workshop of Léon Coulon, who began supplying Cartier in 1884.

17
SI 17 C1900

杯和碟

约 1900 年
卡地亚巴黎
錾刻银镀金
杯　9.15 × 7.13 厘米
碟　15.0 厘米

17
SI 17 C1900

Cup and saucer

Cartier Paris, c. 1900
Chased silver-gilt
Cup 9.15 × 7.13 cm; saucer 15.0 cm

18

HO 08 A02

卷轴式冠冕

1902 年
卡地亚巴黎, 特别订制
银, 金
枕形、圆形旧式切割和玫瑰式切割钻石
种子式镶嵌
中央高度　8.05 厘米

售予埃塞克斯伯爵夫人。

这顶冠冕得以幸存, 十分不易。1990年, 其在伦敦佳士得拍卖公司拍卖, 被一位珠宝商人购得。后者原本想要将冠冕上的宝石拆分下来。在"卡地亚艺术典藏室"前总监埃里克·努斯堡的请求下, 这位善解人意的商人同意将冠冕售予"卡地亚艺术典藏室"收藏, 令这款历史作品得以保存。否则, 它就会和众多其他作品一样, 永远地遗失在历史长河之中。卡地亚档案中现在仍收录了一封埃塞克斯夫人之孙女的感谢信。

18

HO 08 A02

Scroll tiara

Cartier Paris, special order, 1902
Silver, gold
Cushion-shaped, round old- and rose-cut
diamonds
Millegrain setting
Height at centre 8.05 cm

Sold to The Countess of Essex

This tiara only narrowly survived intact. Auctioned at Christie's London in 1990, it was bought by a gem dealer who intended to break it up for the stones. When approached by the former director of the Cartier Collection, Eric Nussbaum, the sympathetic dealer agreed to sell it to the Collection so that this historic item could be preserved. Otherwise, it would have been lost forever, like so many others. The Cartier Archives now has a letter of thanks from the granddaughter of Lady Essex for this effort.

19

BC 15 A02

蝴蝶结纽扣（四颗）

1902 年
卡地亚巴黎
玫瑰金，银
玫瑰式切割钻石
每颗 2.4 × 1.8 厘米

19

BC 15 A02

Four *Bow-knot* buttons

Cartier Paris, 1902
Pink gold, silver
Rose-cut diamonds
Each 2.4 × 1.8 cm

20

CL 114 A03

蕨类植物胸针（两枚）

1903 年
卡地亚巴黎
铂金
圆形旧式切割钻石
种子式镶嵌
18.5 × 4.0 厘米

售予欧内斯特·卡塞尔爵士。

这两款胸针采用全铰接式结构，造型灵活多变。最初的设计是可结合于一个刚性结构上作为冠冕佩戴。1904年，应客人要求对这两款胸针进行了重新镶嵌，可将枝状装饰向外旋转，作为项链或胸饰佩戴。出售的盒子中还搭配了一个扳手型螺丝起子。

欧内斯特·卡塞尔爵士（1852～1921年）是英国国王爱德华七世的朋友兼私人财政顾问。1922年，他的孙女埃德温娜·艾希里嫁给缅甸的蒙巴顿勋爵，后者成为最后一任印度总督。

20

CL 114 A03

Two *fern spray* brooches

Cartier Paris, 1903
Platinum
Round old-cut diamonds
Millegrain setting
18.5 × 4.0 cm

These completely articulated and highly supple brooches could originally be placed on a rigid structure and worn as a tiara. In 1904, the client ordered a setting that made it possible to wear them as a necklace or a corsage ornament by turning the sprays outward. The transformable sprays were delivered in a box with a spanner-head screwdriver.

Sold to Sir Ernest Cassel

Sir Ernest Cassel (1852–1921) was a friend and private financial adviser to King Edward VII of England. In 1922 his granddaughter, the Honorable Edwina Ashley, married the man who would become the last viceroy of India, Earl Mountbatten of Burma.

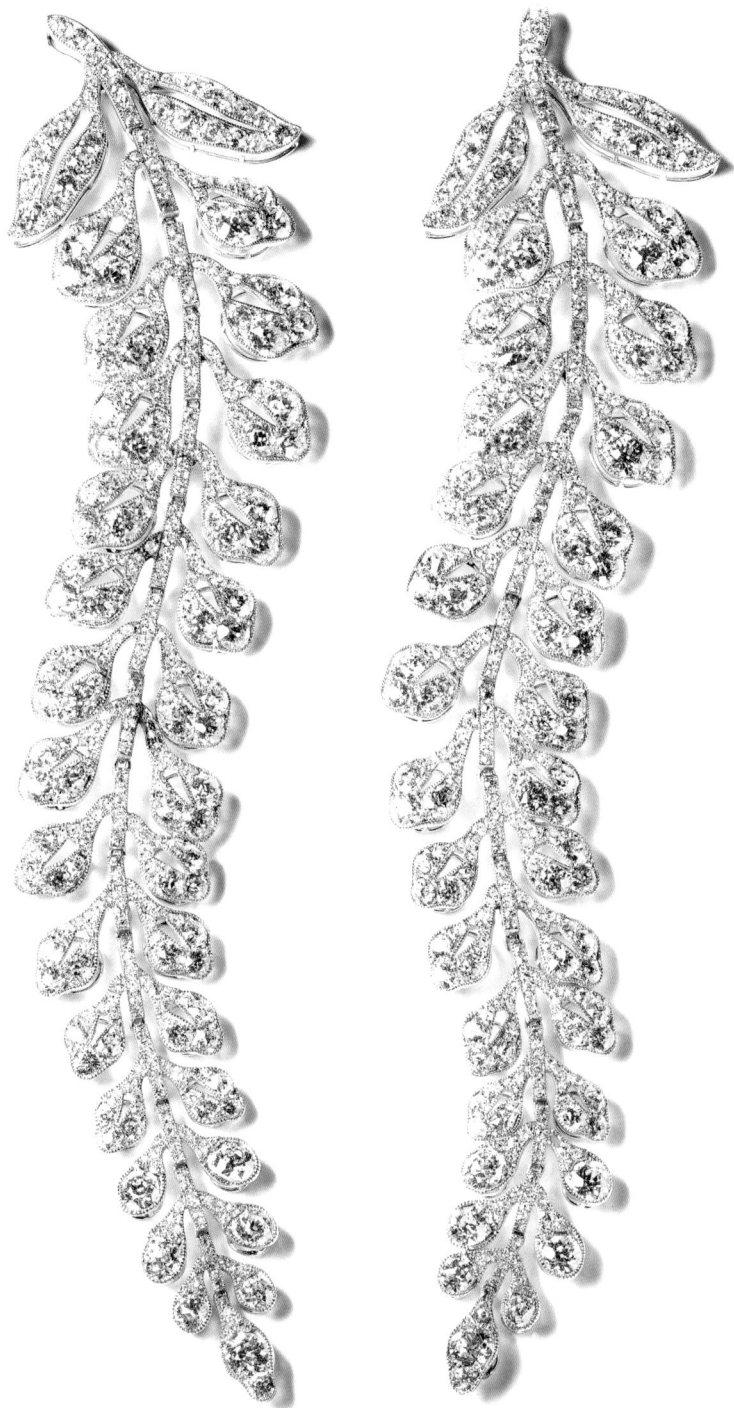

21

CCI 03 A04

带两个环形表盘的座钟

1904 年
卡地亚巴黎
錾刻银镀金
灰蓝色大理石（瓶身）
绿色、蓝色、白色珐琅
高度　24.0 厘米

此座钟为8日储存圆形巴黎机芯，标准擒纵结构。可移动顶盖，内藏指针调节和上链装置。

此座钟是卡地亚1904年所订制的一系列共6款同款式大理石座钟的一款。包括两款灰蓝色大理石，其中一款售予J·P·沃斯；两款角砾大理石，其中一款售予布列特尔侯爵夫人；两款西班牙罗卡角大理石。

21

CCI 03 A04

Clock with 2 annular dials

Cartier Paris, 1904
Chased silver-gilt
Slate-blue marble (vase)
Green, blue and white enamel
Height 24.0 cm

Round 8-day mouvement de Paris, platform escapement. Hand-setting and winding mechanism under the removable cover.

This clock was part of a series of 6 identical marble clocks delivered to Cartier in 1904. Two were done in slate-blue marble, one of which was sold to J. P. Worth; 2 were done in breccia marble (*brèche d'Alep*), one of which was sold to the Marquise de Breteuil; and 2 were in Spanish brocatello marble.

22

CL 92 A05

蝴蝶结胸针

1905 年
卡地亚巴黎
银，金（镀铑）
玫瑰式和欧洲旧式切割钻石
4.0 × 5.1 厘米

售予W·K·范德比尔特夫人。

W·K·范德比尔特夫人原名为安娜·哈里斯曼·山兹·卢瑟福，是卡地亚的忠实客户，于1903年嫁给威廉·吉萨·范德比尔特先生，即康内留斯·范德比尔特之曾孙(参见 CL 06 A25, CL 244 A25, CL 258 A22)。

22

CL 92 A05

Ribbon bow brooch

Cartier Paris, 1905
Silver, gold (rhodium plated)
Rose- and old European-cut diamonds
4.0 × 5.1 cm

Sold to Mrs. W.K. Vanderbilt

Born Anna Harriman Sands Rutherford, this loyal Cartier client married William Kissam Vanderbilt (grandson of "Commodore" Cornelius Vanderbilt) in 1903.

(see CL 06 A25, CL 244 A25, CL 258 A22)

23

PE 20 C05

吊坠

约 1905 年
卡地亚巴黎
金, 银
圆形旧式切割钻石
10.5 × 5.6 厘米

23

PE 20 C05

Pendant

Cartier Paris, c. 1905
Gold, silver
Round old-cut diamonds
10.5 × 5.6 cm

颈饰

1906 年
卡地亚巴黎，特别订制
铂金
圆形旧式切割钻石
种子式镶嵌
33.0 × 5.4 厘米

来源于玛丽·司各特·汤森特夫
人；唐纳德·麦克罗伊夫人（参见 CL
134 A06，NE 16 A10）。

24

NE 39 A06

Choker necklace

Cartier Paris, special order, 1906
Platinum
Round old-cut diamonds
Millegrain setting
33.0 × 5.4 cm

Provenance: Mary Scott Townsend
and Mrs. Donald McElroy
(see CL 134 A06, NE 16 A10)

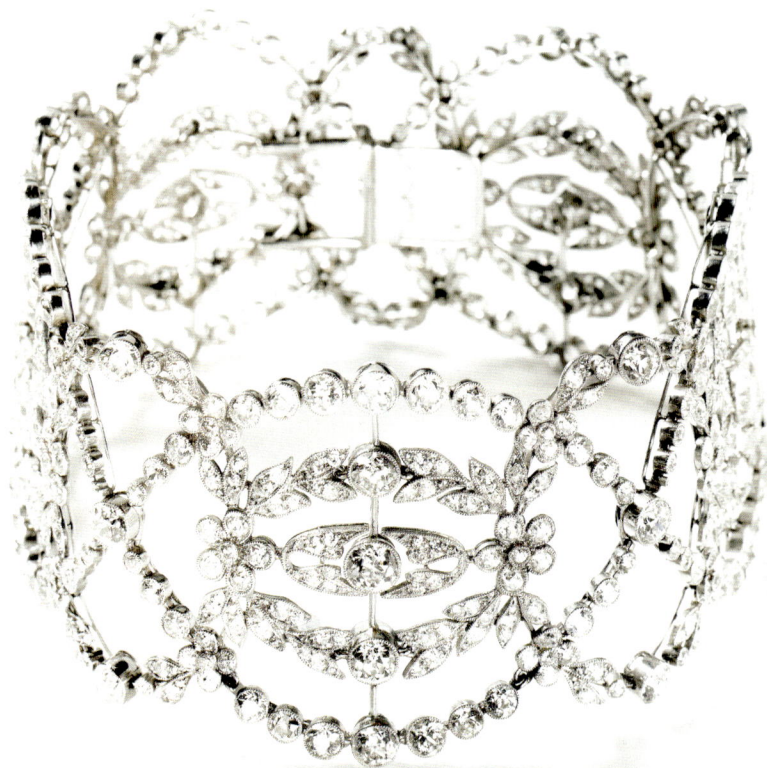

美杜莎头像吊坠

1906 年
卡地亚巴黎
铂金，金
圆形旧式切割和玫瑰式切割钻石
天然珍珠
"天使肌肤"珊瑚浮雕头部和泪滴形坠珠
绿色、白色珐琅
种子式镶嵌
长度　39.5 厘米

美杜莎头像是文艺复兴时期绘画、雕塑和珠宝中常见的经典图式，在19世纪又再度盛行。其原因一方面是因为文艺复兴风格的复苏，另一方面也是出于象征主义和新艺术流派艺术家对这一主题的浓厚兴趣，尤其是与维也纳分离画派有关的艺术家，比如古斯塔夫·克林姆特。这个将美杜莎头像和花环风格镶嵌相结合的吊坠具有路易·卡地亚风格转型时期的典型特征。

25

NE 22 A06

Head of Medusa pendant

Cartier Paris, 1906
Platinum, gold
Round old- and rose-cut diamonds
Natural pearls
"Angel skin" coral cameo head and tear-shaped drop
Green and white enamel
Millegrain setting
Length 39.5 cm

The severed head of Medusa was a motif from classical antiquity that recurred in Renaissance painting, sculpture, and jewellry. It returned again in the nineteenth century thanks not only to the Renaissance-revival style but also to the interest in this subject shown by symbolist and Art-Nouveau artists, notably those associated with the Vienna Secession, such as Gustav Klimt. This pendant, which incorporates a Medusa head in a garland-style setting is characteristic of the stylistic transitions that took place under Louis Cartier.

26

CL 144 A06

蕾丝缎带胸针

1906 年
卡地亚巴黎
铂金, 金
枕形、马眼形、圆形旧式切割和玫瑰式切割钻石
种子式镶嵌
17.1 × 6.45 厘米

售予欧内斯特·卡塞尔爵士（参见
CL 114 A03）。

26

CL 144 A06

Lace ribbon brooch

Cartier Paris, 1906
Platinum, gold
Cushion-shaped, marquise-shaped, round
old- and rose-cut diamonds
Millegrain setting and collet-setting
17.1 × 6.45 cm

Sold to Sir Ernest Cassel
(see CL 114 A03)

百合三角胸衣胸针

1906 年
卡地亚巴黎，特别订制
铂金
圆形旧式切割和玫瑰式切割钻石
种子式镶嵌
每条　27.0 厘米

来源于玛丽·司各特·汤森特；唐纳德·麦克罗伊夫人。

玛丽·司各特·汤森特是 20 世纪初华盛顿上流社会的一名显赫成员。她的侄孙女唐纳德·麦克罗伊（1907～1990年）是司各特—斯特朗煤矿和铁路的继承人。"卡地亚艺术典藏室"还收藏了这两位夫人的另外三件作品：一顶冠冕、一条颈饰（参见 NE 39 A06）、一条项链（参见 NE 16 A10）。

Lily stomacher brooch

Cartier Paris, special order, 1906
Platinum
Round old- and rose-cut diamonds
Millegrain setting
Each branch 27.0 cm

Provenance: Mary Scott Townsend and Mrs. Donald McElroy

Mary Scott Townsend was an eminent member of Washington's high society at the turn of the twentieth century. Her great niece, Thora Ronalds McElroy (1907–1990), was heir to the Scott-Strong coal and railroad fortune. The Collection boasts three other items with the same provenance, a tiara, a choker necklace (see NE 39 A06), and a necklace (see NE 16 A10).

28

HO 26 A07

冠冕

1907 年
卡地亚巴黎，特别订制
铂金
圆形旧式切割钻石
天然珍珠
种子式镶嵌
13.40 × 3.40 厘米

来源于玛丽·波拿巴王妃。

根据档案记载，这款冠冕最初是一个插于脑后的镶嵌橄榄叶形钻石和珍珠的发梳，也可作为束发带佩戴。1907年为玛丽·波拿巴与希腊和丹麦乔治王子的婚礼订制，后改款为冠冕。

玛丽·波拿巴（1882～1962年）是拿破仑之兄吕西安·波拿巴的曾孙女，嫁给希腊乔治一世国王的次子，曾是西格蒙·弗洛伊德的病人、学生、朋友和资助人。她是巴黎精神分析学会（1926年）和法国精神分析协会（1927年）的创始成员。她是弗洛伊德的首位法语翻译，还发表过有关埃德加·艾伦·坡的本能理论和女性性学的文章。卡地亚为她的婚礼所创作的珠宝曾在和平街的卡地亚精品店橱窗内展出。

28

HO 26 A07

Tiara

Cartier Paris, special order, 1907
Platinum
Round old-cut diamonds
Natural pearls
Millegrain setting
13.40 × 3.40 cm

According to the archives, this tiara was originally a comb of diamond olive leaves and pearls for the back of the head, which could also be worn as a bandeau. Made for the wedding of Marie Bonaparte to Prince George of Greece and Denmark in 1907, it was later transformed into a tiara.

Provenance: Princess Marie Bonaparte

Great-granddaughter of Lucien Bonaparte (one of Napoleon's brothers), Marie Bonaparte (1882–1962) married the second son of King George I of Greece before becoming a patient, disciple, friend, and patron of Sigmund Freud. She was a founding member of both the Société Psychanalytique de Paris (1926) and the Revue française de psychanalyse (1927). She was Freud's first French translator and also published writings on Edgar Allen Poe, the theory of instincts, and female sexuality. The jewels that Cartier made for her wedding were displayed in the window of the premises on rue de la Paix.

29

CL 292 A07

三角胸衣胸针

1907 年
卡地亚巴黎，特别订制
铂金
一颗梨形蓝宝石，七颗枕形蓝宝石（总重量为 51
克拉）
圆形旧式切割和玫瑰式切割钻石
21.00 × 12.9 厘米

这款优美的胸针为"卡地亚艺术典藏室"于近期购得，是卡地亚花环风格的典范之作。对于如此大型的一款首饰而言，能够在近100年的时间内依旧保存完好，不得不说是一个奇迹。因为对于大多数冠冕和胸饰来说，一旦款式不再符合潮流，就难逃改款和重镶的命运。

29

CL 292 A07

Stomacher brooch

Cartier Paris, special order, 1907
Platinum
One pear-shaped sapphire
Seven cushion-shaped sapphires
Round old- and rose-cut diamonds
21.00 × 12.9 cm

The total weight of the sapphires is approximately 51 carats.

This beautiful brooch was only recently acquired by the Cartier Collection. It is a splendid example of Cartier's Garland style. It is thoroughly exceptional for a piece of jewellry of this size to have remained completely intact for nearly one hundred years, since transformations and resettings were the inevitable fate of tiaras and corsage ornaments once they went out of fashion.

面纱别针（一对）

1907 年
卡地亚巴黎
金，铂金
玫瑰式切割钻石
圆柱形切割红宝石
3.75 × 1.49 厘米

售予英国国王爱德华七世的妻子亚历珊德拉王后（参见CDS 94 A07）。

来自"斯诺登伯爵夫人玛格丽特公主殿下藏品"。搭配原装卡地亚绿色皮盒。

玛格丽特公主（1930～2002年）是英国国王乔治六世与伊莉莎白王太后的幼女，女王的胞妹（参见CL 296 A38）。

30

JA 30 A07

Pair of veil pins

Cartier Paris, 1907
Gold, platinum
Rose-cut diamonds
Calibré-cut rubies
3.75 × 1.49 cm

Sold to Queen Alexandra, wife of Edward VII, King of England (see CDS 94 A07)

From the Collection of Her Royal Highness the Princess Margaret, Countess of Snowdown. In its original green leather Cartier case.

Princess Margaret (1930-2002) was the younger daughter of King George VI and Queen Elizabeth the Queen Mother, and sister to The Queen.

(see CL 296 A38)

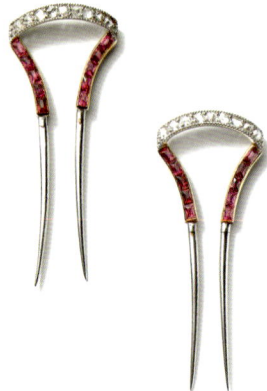

CDS 94 A07

相框座钟

1907 年
卡地亚巴黎
玫瑰金, 银
玫瑰式切割钻石
半透明紫色、白色、蓝色珐琅
凸圆形月长石
10.35 × 6.15 × 7.02 厘米

售予英国国王爱德华七世的夫人
王后亚历珊德拉, 照片为英国王后本
人 (参见 JA 30 A07)。

此座钟为8日储存圆形机芯, 镀
金, 瑞士杠杆式擒纵结构, 双金属平衡
摆轮, 宝玑摆轮游丝。上链和时间调校
轴横穿基座。

放置肖像的位置原本是一个可更
换日历, 共有6页。根据时钟的前主人
介绍, 这款时钟是亚历珊德拉王后赠
送给一位欧洲大使的礼物。这也解释
了王后肖像的由来。

31

CDS 94 A07

Desk clock with
photo frame

Cartier Paris, 1907
Pink gold, silver
Rose-cut diamonds
Translucent mauve enamel over guilloché
ground, white and blue enamel
Moonstone cabochons
10.35 × 6.15 × 7.02 cm

Photo of Queen Alexandra of England

Round 8-day movement, gold-plated, Swiss lever escapement, bimetallic balance, Breguet balance spring. Hand-setting and winding arbor traverses the base.

Originally, the place of the portrait was occupied by a calendar of 6 interchangeable sheets. According to the clock's previous owner, it was given to a European ambassador by Queen Alexandra, which explains the presence of her portrait.

Sold to Queen Alexandra, wife of Edward VII, King of England

(see JA 30 A07)

32

AN 20 C07

长毛犬

约 1907 年
卡地亚巴黎
雕花玛瑙
玫瑰式切割钻石（眼睛）
10.04 × 4.87 厘米

来源于莫拉·罗西·迪·蒙特蕾拉
伯爵夫人，格兰纳德夫人之女（参见
NE 25 A32）。

32

AN 20 C07

Loulou dogs sculpture

Cartier Paris, c. 1907
Carved agate
Rose-cut diamonds (eyes)
10.04 × 4.87 cm

Provenance : Countess Moira Rossi di
Montelera (daughter of Lady Granard,
see NE 25 A32)

33

WB 36 A08

表胸针

1908 年
卡地亚巴黎
铂金，金
半透明铁灰色、白色珐琅
珍珠，一颗切面祖母绿，圆柱形切割祖母绿
玫瑰式切割钻石
长度　9.9 厘米

售予英国国王爱德华七世。

此表为积家10HPVB型圆形机芯，镀金，18颗宝石轴承，瑞士杠杆式擒纵结构，双金属平衡摆轮，扁平摆轮游丝。

爱德华七世（1841～1910年）是维多利亚女王之子，亦是英格兰和爱尔兰国王，1901～1910年期间在位。他赞誉卡地亚为"皇帝的珠宝商，珠宝商的皇帝"。

33

WB 36 A08

Watch-brooch

Cartier Paris, 1908
Platinum, gold
Translucent iron-gray enamel over guilloché
ground, white enamel
Pearls, one faceted emerald, calibré-cut
emeralds
Rose-cut diamonds
Length 9.9 cm

Round LeCoultre caliber 10HPVB movement, gold-plated, 18 jewels, Swiss lever escapement, bimetallic balance, flat balance spring.

Sold to Edward VII, King of England

Edward VII (1841-1910) was the son of Queen Victoria, and King of England and Ireland from 1901 to 1910. He proclaimed Cartier as "Jeweller of Kings, King of Jewellers."

34

BS 03 A08

椭圆形糖果盒

1908 年
卡地亚巴黎
金
玫瑰色石英
白色珐琅
4.91 × 3.25 × 2.43 厘米

售予齐妮亚大公夫人。

34

BS 03 A08

Oval box bonbonnière

Cartier Paris, 1908
Gold
Rose quartz
White enamel
4.91 × 3.25 × 2.43 cm

Sold to the Grand Duchess Xenia

带温度计的座钟

1908 年
卡地亚巴黎
铂金，金，银，银镀金
玫瑰式切割钻石
半透明紫色珐琅
17.3 × 8.1 × 6.4 厘米

售予葡萄牙国王。

此钟为8日储存圆形机芯，镀金，瑞士杠杆式擒纵结构，双金属平衡摆轮，宝玑摆轮游丝。时间调校按钮位于指针中央；上链轴位于6点钟位置。

35

CDS 49 A08

Desk clock with thermometer

Cartier Paris, 1908
Platinum, gold, silver, silver-gilt
Rose-cut diamonds
Translucent mauve enamel over guilloché ground
17.3 × 8.1 × 6.4 cm

Round 8-day movement, gold-plated, Swiss lever escapement, bimetallic balance, Breguet balance spring. Hand-setting knob at center of hands; winding arbor at 6 o'clock.

Sold to the King of Portugal

花篮式吊坠表

1909 年
卡地亚巴黎
铂金，玫瑰金
旧式切割和玫瑰式切割钻石
灰蓝色半透明珐琅
链绳长度　69.0 厘米
表　3.3 × 3.1 厘米

售予女高音歌唱家蕾丽·梅尔巴，后于1912年转售予尤索波夫亲王。

按压花篮正中的钻石，即可打开中心的花朵，读取表盘。

此表为积家9HPVB型圆形机芯，镀金，18颗宝石轴承，瑞士杠杆式擒纵结构，双金属平衡摆轮，扁平摆轮游丝。

尤索波夫家族极其富有。菲利克斯·尤索波夫亲王（1887～1967年）于1914年迎娶沙皇尼古拉斯二世的侄女爱丽娜·亚历山德拉多娃女大公。1916年，亲王成为拉斯普京暗杀案的主策划人之一。在俄国大革命之后，他离开俄国前往法国，随身携带了部分瑰丽的家族珠宝（参见 CL 156 A13）。

Floral basket pendant watch

Cartier Paris, 1909
Platinum, pink gold
Old- and rose-cut diamonds
Gray-blue translucent enamel
Length of the chain 69.0 cm; watch 3.3 × 3.1 cm

Pressure on the diamond in the middle of the basket opens the central flower to reveal the dial.

Round LeCoultre caliber 9HPVB movement, gold-plated, 18 jewels, Swiss lever escapement, bimetallic balance, flat balance spring.

Sold to the soprano Nellie Melba, then to Prince Yusupov in 1912

The Yusupov family was immensely rich. In 1914, Prince Felix Yusupov (1887–1967) married Grand Duchess Irina Alexandrovna, niece of Tsar Nicholas II. In 1916, the prince was one of the main plotters behind the assassination of Rasputin. After the Revolution he left Russia for France, taking with him some of the magnificent family jewels.

(see CL 156 A13)

蝴蝶结胸针

1910 年
卡地亚巴黎
铂金
圆形旧式切割和玫瑰式切割钻石
天然珍珠
种子式镶嵌
2.5 × 6.2 厘米

Bow brooch

Cartier Paris, 1910
Platinum
Round old- and rose-cut diamonds
Natural pearls
Millegrain setting
2.5 × 6.2 cm

38

HO 02 A10

卷轴式冠冕

1910 年
卡地亚巴黎
铂金
一颗枕形钻石, 圆形旧式切割钻石
种子式镶嵌
中央高度　5.5 厘米

售予比利时王后伊丽莎白
(1875～1965年)。

伊丽莎白嫁给阿尔伯特一世亲
王。亲王于1909年即位, 1934年驾崩。

38

HO 02 A10

Scroll tiara

Cartier Paris, 1910
Platinum
One cushion-shaped diamond, round old-
cut diamonds
Millegrain setting
Height at centre 5.5 cm

Sold to Elisabeth, Queen of the Bel-
gians (1875-1965)

She married Prince Albert 1st who
became King in 1909 and died in
1934.

39

PF 01 C10

相框

约 1910 年
卡地亚巴黎
金，银
玉
白色珐琅
直径 18.8 厘米

画像为弗拉基米尔大公之妻玛丽亚·巴甫洛娃的肖像。

39

PF 01 C10

Photo frame

Cartier Paris, c. 1910
Gold, silver
Nephrite
White enamel
Diameter 18.8 cm

Portrait of Maria Pavlovna, wife of the Grand Duke Vladimir.

40

NE 05 A11

项链

1911 年
卡地亚巴黎，特别订制
铂金
圆形旧式切割和玫瑰式切割钻石
天然珍珠
种子式镶嵌
长度　45 厘米

珍珠重量（1格令 = 0.05 克）：
一颗粉灰色梨形珍珠重107.60格
令；一颗粉灰色纽扣形珍珠重88.16格
令；三颗灰色梨形珍珠分别重64.28格
令、59.48格令、55.92格令；一颗灰色
圆形珍珠重12.5格令。

40

NE 05 A11

Necklace

Cartier Paris, special order, 1911
Platinum
Round old- and rose-cut diamonds
Natural pearls
Millegrain setting
Length 45 cm

Weight of the pearls (1 grain = 0.05 gram):

one pinkish-gray pear-shaped pearl: 107.60 grains. One pinkish-gray button pearl: 88.16 grains. Three gray pear-shaped pearls: 64.28, 59.48 and 55.92 grains. One gray round pearl: 12.5 grains.

Each of the pearls is certified.

NE 11 A12

吊坠

1912 年
卡地亚巴黎
铂金
圆形旧式切割和玫瑰式切割钻石
两颗凸圆形星彩蓝宝石（分别重 13.45 克拉和
1.98 克拉，附有鉴定证书，证明两颗星彩蓝宝石
均为天然宝石）
垂坠圆形天然珍珠
雕花无色水晶
种子式镶嵌
11.5 × 4.5 厘米

来源于史蒂芬·西尔维先生和夫
人。

此吊坠中间部分可旋下，作为胸
针佩戴。吊坠两侧有两个雕花无色水
晶半人马头。该杰作是卡地亚最早的
无色水晶作品之一。

41

NE 11 A12

Pendant

Cartier Paris, 1912
Platinum
Round old- and rose-cut diamonds
Two star sapphire cabochons (13.45 and
1.98 carats)
Round and pendant natural pearls
Carved rock crystal
Millegrain setting
11.5 × 4.5 cm

A certificate confirms that both star
sapphires are natural.

The central section can be un-
screwed and worn as a brooch. Two
faun heads carved in rock crystal can
be seen in profile on either side of
the pendant. This outstanding piece
is one of Cartier's first works in rock
crystal.

Provenance: Mr. and Mrs. Stephen
Silver

42

CL 122 A12

三角胸衣胸针

1912 年
卡地亚巴黎，特别订制
铂金
圆形旧式切割和玫瑰式切割钻石
种子式镶嵌
11.0 × 9.50 厘米

42

CL 122 A12

Stomacher brooch

Cartier Paris, special order, 1912
Platinum
Round old- and rose-cut diamonds
Millegrain setting
11.0 × 9.50 cm

43

CL 283 A12

蝴蝶结胸针

1912 年
卡地亚巴黎
铂金
一颗圆形旧式切割钻石，单面切割钻石
雕花无色水晶
5.73 × 1.80 厘米

43

CL 283 A12

Bow brooch

Cartier Paris, 1912
Platinum
One round old-cut diamond, single-cut
diamonds
Carved rock crystal
5.73 × 1.80 cm

44

JA 18 A12

帽针

1912 年
卡地亚巴黎
铂金
明亮式切割钻石
雕花无色水晶
长度 12.0 厘米

44

JA 18 A12

Hat pin

Cartier Paris, 1912
Platinum
Brilliant-cut diamonds
Carved rock crystal
Length 12.0 cm

CL 105 A13

吊坠胸针

1913 年
卡地亚巴黎，特别订制
铂金，金
五颗枕形蓝宝石（总重 39.66 克拉）
切面和圆柱形切割蓝宝石
圆形旧式切割钻石
种子式镶嵌
12.09 × 6.02 × 0.78 厘米

来源于埃斯特哈兹伯爵夫人玛丽 - 艾米莉，即约瑟夫 - 纪尧姆•帕尔夫•道恩之妻。

这枚胸针乃是由客户所提供的一条项链和吊坠改款而成。类似的带有两个不同元素吊坠的胸针在卡地亚被称为"吊坠胸针"（源自花环风格的"垂坠"效果）。

45

CL 105 A13

Brooch-pendant

Cartier Paris, special order, 1913
Platinum, gold
Five cushion-shaped sapphires, weighing
39.66 carats in total
Faceted and calibré-cut sapphires
Round old-cut diamonds
Millegrain setting and collet-setting
12.09 × 6.02 × 0.78 cm

This brooch was made from two earlier pieces, a necklace and a pendant, supplied by the client. Such brooches, with two distinct elements from which the pendant hung, were called "drapery brooches" by Cartier (from the "drape" effect of garland-style jewellry).

Provenance: Marie-Emilie, Countess Esterházy (wife of Josèphe-Guillaume Palffy Daun)

46

CL 156 A13

面纱别针

1913 年
卡地亚巴黎
铂金
圆形旧式切割和玫瑰式切割钻石
一颗天然珍珠
种子式镶嵌
2.39 × 8.35 厘米

售予尤索波夫亲王（参见 WN 11 A09）。

46

CL 156 A13

Veil pin

Cartier Paris, 1913
Platinum
Round old- and rose-cut diamonds
One natural pearl
Millegrain setting
2.39 × 8.35 cm

Sold to Prince Yusupov

(see WN 11 A09)

47

HO 11 A14

冠冕

1914 年
卡地亚巴黎，特别订制
黑钢、铂金
梨形、圆形旧式切割和玫瑰式切割钻石
切面和圆柱形切割红宝石
中央高度　4.1 厘米

据档案记载，这顶冠冕源自1906
年制作，后售予艾佛瑞夫人的一款项
链。1914年，卡地亚重新利用项链上的
9颗梨形钻石和四周环绕的圆柱形切
割红宝石，制作出了这顶冠冕。客人是
罗马尼亚玛丽王后的友人，可能是为
王后与费迪南德国王的加冕礼制作。
在这一时期，卡地亚还制作了数顶类
似的钢质冠冕。

47

HO 11 A14

Tiara

Cartier Paris, special order, 1914
Blackened steel, platinum
Pear-shaped, round old- and rose-cut
diamonds
Faceted and calibré-cut rubies
Height at centre 4.1 cm

According to the archives, this tiara
derived from a necklace originally
made for stock in 1906 and sold to
Lady Avery. In 1914, nine pear-shaped
diamonds and the surrounding
calibré-cut rubies were reused for
this tiara. The client was a friend of
Queen Mary of Romania and the tiara
was probably made on the occasion
of the queen's coronation alongside
King Ferdinand. Cartier produced
several steel tiaras of this type at the
time.

48

CM 19 A14

Model A 魅幻时钟

1914 年
卡地亚巴黎
金，铂金
无色水晶，玛瑙（基座）
四颗凸圆形蓝宝石，玫瑰式切割钻石
白色珐琅
高度 13.0 厘米

售予格雷夫尔伯爵。

此钟为8日储存长方形机芯，镀金，瑞士杠杆式擒纵结构，双金属平衡摆轮，宝玑摆轮游丝。时间调校和上链轴位于基座下方。

第一款Model A 魅幻时钟由卡地亚于1912年售出。魅幻时钟之所以神秘，是因为其铂金镶钻指针看上去好像没有和任何机械机芯相连。事实上，每一枚指针都镶嵌在一个带有隐藏齿缘的扁平水晶盘上；齿缘由隐藏在时钟两侧的垂直支架驱动。而这两个支架本身则由位于基座的机芯所驱动。

格雷夫尔伯爵为著名的格雷夫尔伯爵夫人的丈夫。格雷夫尔伯爵夫人被马歇尔·普鲁斯特誉为"欧洲最美丽的女人"，是普鲁斯特笔下的盖尔芒特公爵夫人的原型之一。

48

CM 19 A14

Model A mystery clock

Cartier Paris, 1914
Gold, platinum
Rock crystal, agate (base)
Four cabochon sapphires, rose-cut
diamonds
White enamel
Height 13.0 cm

Rectangular, 8-day movement, gilded, Swiss lever escapement, bimetallic balance, Breguet balance spring. Hand-setting and winding mechanism underneath the base.

Sold to Count Greffulhe

Husband to the famous Countess Greffulhe, "the most beautiful women in Europe" according to Marcel Proust, who partly modeled his character of the Duchess de Guermantes on her.

The very first Model A mystery clock was sold by Cartier in 1912.

They are mysterious because their platinum and diamond hands do not seem to be linked to any mechanical movement. In fact each hand of the clock is set on a flat disk of crystal with a hidden, toothed edge; the toothed disks are driven by two vertical racks hidden in the sides of the clock; these racks are themselves driven by the movement in the base.

49

CL 256 A18

花篮胸饰

1918 年
卡地亚纽约
铂金
六角形、半月形、方形、圆形旧式切割和单面切
割钻石
天然珍珠
花篮采用种子式镶嵌
9.45 × 7.40 × 1.60 厘米

这枚胸针是后期在美国所兴起的
不对称花环风格的代表作。

49

CL 256 A18

Flower basket corsage

ornament

Cartier New York, 1918
Platinum
Hexagonal, half-moon, square, round old-
and single-cut diamonds
Natural pearls
Millegrain setting for the basket
9.45 × 7.40 × 1.60 cm

This brooch provides an example of
the later, asymmetrical garland style
developed in America.

50

RG 03 A20

戒指

1920 年
卡地亚伦敦
铂金
一颗枕形蓝宝石（重 18.87 克拉）
圆形旧式切割和单面切割钻石
2.76 × 1.78 × 1.55 厘米

50

RG 03 A20

Ring

Cartier London, 1920
Platinum
One cushion-shaped sapphire weighing
18.87 carats
Round old- and single-cut diamond
2.76 × 1.78 × 1.55 cm

51
NE 08 A21

吊坠

1921 年
卡地亚巴黎，特别订制
铂金
圆形旧式切割、单面切割和玫瑰式切割钻石
一颗凸圆形祖母绿和两颗水滴形祖母绿
雕花无色水晶
两颗黑玛瑙饰珠
黑色珐琅扣环
长度　11.30 厘米

这个吊坠乃是从1913年的一个款式特别改款而来。这说明该风格在1921年已经过时。

51
NE 08 A21

Pendant

Cartier Paris, special order, 1921
Platinum
Round old-, single- and rose-cut diamonds
One emerald cabochon and two drop-shaped emeralds
Carved rock crystal
Two onyx beads
Black enamel clasp
Length 11.30 cm

This pendant was specially commissioned from a 1913 model, which explains the use of a style that had become outmoded by 1921.

化妆盒

约 1924 年
卡地亚纽约
金，铂金
金色和白色条纹珐琅装饰，白色、黑色珐琅
玫瑰式切割钻石
8.40 × 4.05 × 3.00 厘米

售予威斯敏斯特公爵。

顶盖镌刻"爱你的本多尔，24"。
内设三个翻盖隔层，其中，一个为唇膏
架，一个为香烟匣。

据说这个化妆盒由威斯敏斯
特公爵于1924年赠送给他的情人，
著名法国时装设计师可可·香奈儿
（1883～1971年）。

5²

VC 39 C24

Vanity case

Cartier New York, c. 1924
Gold, platinum
Striped gold-and-white enamel decoration
(*pékin*), white and black enamel
Rose-cut diamonds
8.40 × 4.05 × 3.00 cm

The lid engraved:"Amour Ben d'or '24"
The interior fitted with three lidded
compartments, one containing a
lipstick holder and one for the cigarettes.

Sold to the Duke of Westminster

This vanity case is said to have
been given in 1924 by the Duke of
Westminster to his mistress Coco
Chanel (1883-1971), the famous
French couturiere.

53

HO 12 A37

冠冕

1937 年
卡地亚伦敦
铂金
圆形旧式切割钻石
椭圆形和花式切割海蓝宝石
中央高度 5.04 厘米

　　中央的饰物可以从冠冕上取下，用作胸针（头向下）。

　　最初这个冠冕是由单排椭圆形海蓝宝石组成，5个月后，应客人要求，又加了一排同样的宝石。1937年卡地亚伦敦接到不少于27个冠冕的订单，其中大多数都是用于英国国王乔治六世在同年5月的加冕仪式上佩戴。

53

HO 12 A37

Tiara

Cartier London, 1937
Platinum
Round old-cut diamonds
Oval and fancy-cut aquamarines
Height at centre 5.04 cm

The central motif can be removed from the tiara and worn as a brooch, pointing downward.

This tiara was first made with a single row of oval aquamarines, but the client had a second row of identical stones added five months later. In 1937 Cartier London received orders for no fewer than twenty-seven tiaras, most of them to be worn at the coronation of King George VI in May of that year.

CL 296 A38

玫瑰胸针

1938 年
卡地亚伦敦
铂金
长阶梯形切割、圆形旧式切割和单面切割钻石
7.32 × 4.32 厘米

来源于玛格丽特公主殿下，即斯诺登伯爵夫人（参见JA 30 A07）。

1953年6月2日，在威斯敏斯特教堂，玛格丽特公主殿下于其姐姐伊丽莎白二世陛下的加冕礼上佩戴此胸针（参见2006年6月13日伦敦佳士得目录）。

54

CL 296 A38

Rose clip brooch

Cartier London, 1938
Platinum
Baguette-, round old- and single-cut
diamonds
7.32 × 4.32 cm

Provenance: HRH The Princess Margaret, Countess of Snowdon (see JA 30 A07)

H.R.H. The Princess Margaret wore this brooch to the coronation of her sister H.M. Queen Elizabeth II on 2nd June 1953 at Westminster Abbey. (see catalogue Christie's London, June 13, 2006)

55

BT 116 A39

花朵手镯

1939 年

卡地亚巴黎, 特别订制

铂金, 白金

一颗枕形钻石（重约 4.50 克拉），明亮式切割和
圆形旧式切割钻石

胸针　6.00 厘米

手镯直径　5.00 厘米

售予一名罗马尼亚皇室成员。

55

BT 116 A39

Bangle with *Flower*

Cartier Paris, special order, 1939
Platinum, white gold
One cushion-shaped diamond weighing
approximately 4.50 carat, brilliant- and
round old-cut diamonds
6.00 cm (brooch); 5.00 cm (diameter of the
bracelet)

Sold to a member of the Royal family
of Romania

56

RG 33 A47

戒指

1947 年

卡地亚巴黎, 特别订制

圆模雕刻装饰金和金丝, 铂金

明亮式切割钻石

方形切割面祖母绿

一颗凸圆形珊瑚

3.80 × 3.00 × 2.90 厘米

售予温莎公爵夫人（参见 NE 09
A47, CL 127 A53, CL 251 A55, CL 53
A49, OI 08 A54）。

56

RG 33 A47

Ring

Cartier Paris, special order, 1947
Gadrooned gold and gold wire, platinum
Brilliant-cut diamonds
Square faceted emeralds
One coral cabochon
3.80 × 3.00 × 2.90 cm

Sold to the Duchess of Windsor
(see NE 09 A47, CL 127 A53, CL
251 A55, CL 53 A49, OI 08 A54)

57

NE 09 A47

吊坠项链

1947 年
卡地亚巴黎，特别订制
螺旋 18K 和 20K 金，铂金
明亮式切割和长阶梯形切割钻石
一颗心形切面紫水晶，27 颗祖母绿式切割紫水
晶，一颗椭圆形切面紫水晶
凸圆形绿松石
20.0 × 19.5 厘米

此款项链售予温莎公爵。除了绿
松石外，其他宝石均为温莎公爵提供。

来源于温莎公爵夫人。

沃利斯·沃菲尔德（1896～1986
年）生于巴尔的摩，首任丈夫是温菲尔
德·斯宾塞伯爵，之后嫁给了美国富商
恩斯特·辛普森。接着在1934年，她遇
见了威尔斯王子。他们的恋情在大英
帝国引起了轩然大波。威尔斯王子于
1936年即位，而沃利斯也结束了她的
第二段婚姻。同年的12月，爱德华八
世为了迎娶沃利斯，不得不选择退位。
1937年3月，新的国王乔治六世，封
其兄弟为温莎公爵。6月，两人在法国
完婚，并移居巴黎，且长久定居于此。
自20世纪30年代后期，公爵夫人开始
迷恋上卡地亚。她的很多珠宝都是与
贞·杜桑女士合作完成的（参见RG 33
A47, CL 127 A53, CL 251 A55, CL 53
A49, OI 08 A54）。

57

NE 09 A47

Bib necklace

Cartier Paris, special order, 1947
Twisted 18-carat and 20-carat gold,
platinum
Brilliant- and baguette-cut diamonds
One heart-shaped faceted amethyst,
twenty-seven emerald-cut amethysts, one
oval faceted amethyst
Turquoise cabochons
20.0 × 19.5 cm

This necklace was sold to the Duke of
Windsor, who supplied all the stones
except the turquoises.

Provenance: the Duchess of Windsor

Born in Baltimore, Wallis Warfield
(1896–1986) first married Earl Winfield
Spencer, followed by a rich American
businessman, Ernest Simpson. Then,
in 1934, she met the Prince of Wales.
Their affair caused a scandal in Great
Britain, and following his accession to
the throne in 1936 and Wallis's second
divorce, Edward VIII was forced to ab-
dicate in December in order to marry
her. In March 1937 the new king,
George VI, named his brother Duke
of Windsor. The couple were married
in France in June, and moved to Paris
whether they lived for the rest of their
lives. From the late 1930s onward the
Duchess was a great fan of Cartier.
Many of her jewels were made in col-
laboration with Jeanne Toussaint.

(see RG 33 A47, CL 127 A53, CL
251 A55, CL 53 A49, OI 08 A54)

58

CL 53 A49

猎豹胸针

1949 年
卡地亚巴黎
铂金，白金
单面切割钻石
两颗梨形黄色钻石（眼睛）
一颗克什米尔凸圆形蓝宝石（重 152.35 克拉）
凸圆形蓝宝石（斑点）
6.0 × 3.7 × 3.0 厘米

售予温莎公爵夫人（参见 NE 09 A47, RG 33 A47, CL 127 A53, CL 251 A55, OI 08 A54）。

58

CL 53 A49

Panther clip brooch

Cartier Paris, 1949
Platinum, white gold
Single-cut diamonds
Two pear-shaped yellow diamonds. (eyes)
One 152.35-carat Kashmir sapphire cabochon
Sapphire cabochons (spots)
6.0 × 3.7 × 3.0 cm

Sold to the Duchess of Windsor
(see NE 09 A47, RG 33 A47, CL 127 A53, CL 251 A55, OI 08 A54)

59

CL 127 A53

鸭头胸针

1953 年
卡地亚巴黎
金，铂金
一颗明亮式切割钻石
圆形切面祖母绿
一颗椭圆形切面蓝宝石
附壳珍珠
雕花珊瑚
4.2 × 4.8 厘米

售予温莎公爵夫人（参见 NE 09 A47, RG 33 A47, CL 251 A55, CL 53 A49, OI 08 A54）。

59

CL 127 A53

Duck's head clip brooch

Cartier Paris, 1953
Gold, platinum
One brilliant-cut diamond
Round faceted emeralds
One oval faceted sapphire
Blister pearl
Carved coral
4.2 × 4.8 cm

Sold to the Duchess of Windsor
(see NE 09 A47, RG 33 A47, CL 251 A55, CL 53 A49, OI 08 A54)

60

OI 08 A54

虎形长柄眼镜

1954 年
卡地亚巴黎, 特别订制
金
两颗马眼形祖母绿(眼睛)
内嵌黑色珐琅(虎纹)
长度 8.5 厘米(合拢状态)

铰链式折叠长柄眼镜藏于老虎体内。

温莎公爵夫人特别订制(参见NE 09 A47, RG 33 A47, CL 127 A53, CL 251 A55, CL 53 A49)。

60

OI 08 A54

Tiger lorgnette

Cartier Paris, special order, 1954
Gold
Two navette-shaped emeralds (eyes)
Champlevé black enamel (stripes)
Length 8.5 cm (closed)

The hinged fold out lorgnette concealed behind the tiger's body.

Made to special order for the Duchess of Windsor.

(see NE 09 A47, RG 33 A47, CL 127 A53, CL 251 A55, CL 53 A49)

61

CL 251 A55

哈巴狗头胸针

1955 年
卡地亚巴黎
金
两颗凸圆形黄水晶(眼睛)
土黄色、棕色、黑色和白色珐琅(头部)
红色、蓝色、白色和绿色景泰蓝(领子)
2.4×3.3×3.2 厘米

售予温莎公爵(参见NE 09 A47, RG 33 A47, CL 127 A53, CL 53 A49, OI 08 A54)。

61

CL 251 A55

Pug's head clip-brooch

Cartier Paris, 1955
Gold
Two citrine cabochons (eyes)
Ochre-coloured, brown, black and white enamel (head)
Cloisonné enamelled red, blue, white and green (collar)
2.4×3.3×3.2 cm

Sold to the Duke of Windsor

(see NE 09 A47, RG 33 A47, CL 127 A53, CL 53 A49, OI 08 A54)

尽管路易·卡地亚偏爱 18 世纪的设计风格，并在 20 世纪初以花环风格和堪与俄罗斯法贝热彩蛋媲美的精美珐琅配饰 (参见 WB 36 A08, DI 24 A08, CCI 04 A07) 大获成功，他仍鼓励设计师们去探寻更加现代的美学形式，打破路易十六风格的繁复华丽。因此，卡地亚工作坊开始创作几何线条（方形、圆形、菱形）的胸针，并在铂金和钻石的单色世界中，引入了缤纷的红宝石和蓝宝石作为装饰 (参见 CL 288 A04, CL 300 A08)，同时还将祖母绿与蓝宝石、紫水晶与红宝石组合在一起，甚至尝试其他更为大胆的色彩搭配。此后，现代主义的清新气息无处不在。在 1914 年创作的一顶冠冕中，卡地亚甚至采用了黑钢，与铂金、红宝石和钻石互相搭配 (参见 HO 11 A14)。

一战前夕，卡地亚开始采用一种特殊的材质——黑玛瑙。高度抛光的黑玛瑙被用作微妙的阴影线，创造出神奇的径深效果，令钻石更加晶莹明亮，珠宝和腕表的线条更为清晰明朗 (参见 NE 01 A13)，为珠宝赋予了生动的立体感 (参见 CL 138 A22)，成为卡地亚偏爱的材质之一。早在 1914 年，"豹纹"图案就已出现在腕表上。它以黑玛瑙和钻石组成，后来成为了卡地亚的标志性图案之一，并一直沿用至今。1917 年，猎豹造型首次亮相于一件名贵的化妆盒 (参见后期的一件作品 VC 08 A28)，与真正的豹纹一样，每一个斑点都各不相同，这意味着每一颗宝石都必须采用不同的切割手法，镶嵌于精致的基座之内，点缀在爪式铺镶的钻石之间。除钻石外，黑玛瑙还可与几乎所有名贵宝石，尤其是珊瑚搭配 (参见 HO 24 A22)，因此珊瑚也成为卡地亚最青睐的材质之一。卡地亚使用的是一种深色珊瑚，更偏向橙色调，而非红色或粉色，这种珊瑚与深绿色绿宝石搭配，创造出极富原创性的珠宝 (参见 BT 44 A21)，并主要用于东方风格的设计之中。

但路易·卡地亚最心仪的色彩组合无疑是蓝色和绿色。青金石和绿松石出现在手链和造型优美的烟盒上，虽然也曾采用玉和绿松石 (参见 CL 193 A13) 的组合，但是最名贵的则莫过于蓝宝石和祖母绿的组合。1923 年，卡地亚创作了一件名贵华丽的项坠 (参见 NE 24 A23)，以一颗重达 121 克拉的凸圆蓝宝石搭配雕花祖母绿。这组蓝色与绿色的组合，在抛光与雕花表面呈现出微妙的肌理变化，创造出轻快和谐的蓝色韵律。20 世纪 30 年代，时尚之风向柔美转变。波浪长发开始流行。服装的剪裁也开始注重身体的曲线，腰线再次变得明朗。珠宝的设计也随之发生了变革：由 20 世纪 20 年代的扁平几何设计，逐渐演变为三维立体作品——柔和的弧线造型，或富有建筑感的设计。

1933 年，贞·杜桑女士被指派为卡地亚的高级珠宝总监，负责引领整个珠宝工作坊的设计，维持卡地亚的完美标准。即便在大萧条和二战的黑暗年代中，她也充分发挥了丰富的想象力和智慧，开创了一种全新的自然主义风格，动物和植物成为卡地亚奇妙幻想世界的一部分。在这一时期，白色金属尊崇不再，黄金卷土重来。此外，卡地亚还推出了新奇而大胆的色彩组合：以黄水晶搭配紫水晶（参见 BT 105 A39），或是紫水晶搭配海蓝宝石（参见 CL 132 A37）。

卡地亚设计了大量精美的镶钻花卉，其中两枚成为了英国女王伊丽莎白及其妹妹玛格丽特公主殿下的最爱（参见 CL 296 A38）。此外，珊瑚一直以来也是卡地亚作品中的常见材质，为此，设计师们还独创了一种仿生式镶嵌法，尽可能地从自然界汲取灵感。这一系列镶嵌法于 1968 年投入实践，当时，野性不羁的墨西哥女星玛莉亚·菲利克斯向卡地亚提出了设计一条蛇形项链的挑战（参见 NE 10 A68）。巴黎工作坊特别为她创作了一款令人叹为观止的艺术作品，以 2473 颗钻石组成，总长 57 厘米，重 600 克，整条项链看上去栩栩如生。数年后，卡地亚再度应邀设计制作了由两条鳄鱼组成的绝妙作品，镶嵌了数千颗黄钻石和祖母绿，既可作为两枚胸针分别佩戴；又可将两条鳄鱼首尾交叉，作为一条项链佩戴（参见 NE 43 A75），堪称珠宝史上最为独特的作品。

卡地亚艺术典藏室馆长　帕斯卡尔·勒博

Despite his taste for the 18th century and the extraordinary success of the Garland style at the climax of its popularity in the beginning of the 20th century, together with delicate enameled accessories equaling those of the Russian Jeweller Fabergé (See WB 36 A08, DI 24 A08, CCI 04 A07), Louis encouraged his designers to strive for a more modern aesthetic approach, one that broke away from the exuberance of the Louis XVI style. Accordingly, the workshops began creating brooches with geometric lines – squares, circles, lozenges– introducing the colors of ruby and sapphire into the previously monochrome universe of platinum and diamonds (See CL 288 A04, CL 300 A08). Such flamboyant juxtapositions confirmed the possibility of combining emeralds with sapphires, amethysts with rubies, and other bold color combinations. Their fervor of modernity was such that Cartier would even employ blackened steel in combination with platinum, rubies, and diamonds as seen in an incredible tiara created in 1914 (see HO 11 A14).

The pre-war period was marked by the use of a material that would become a Cartier fetish: onyx. Highly polished, dark black onyx enhanced diamonds, and reinforced the lines of jewels and watches (see NE 01 A13). Subliminally used as a shadow line, onyx created the illusion of depth, adding an additional third dimension to jewels (see CL 138 A22). The "panther-skin" motif, combining onyx and diamonds was created in 1914, initially used on a wristwatch. Still used to this day, it has become one of Cartier's classic motifs. It was in 1917 that the panther made its first figurative appearance on a precious vanity case (see VC 08 A28 for a later example); Just like real panther skin, each spot is different, thus each stone had to be cut uniquely, then held by a delicate millegrain setting within a pavé ground of claw-set diamonds. Onyx was not only combined with diamonds but with all other precious stones, especially with coral (see HO 24 A22), which also became one of the firms' favorite materials. Cartier used a deep-hued coral, more orange than it was red or pink. Its combination with deep green emeralds gave birth to highly original jewels (see BT 44 A21), and it was largely employed in designs of an Oriental influence.

Yet, Louis Cartier's favorite color combination was, without any doubt, blue and green. Lapis lazuli and turquoise could be found both from bracelet and elegant shaped cigarette case. Jade and turquoise (see CL 193 A13) could also be combined, however the most precious gem combination was that of sapphires with emeralds. In 1923, a magnificent pendant (see NE 24 A23) composed of a sapphire cabochon, weighing 121 carats, and engraved emeralds was created, blending the delightful blend of colors with a subtle play of polished and carved surfaces. In the 1930's fashion became more feminine, hair was worn longer and with waves, and dresses followed the body curves, with the waist once again becoming apparent. Jewels changed accordingly, and the flat geometric design of the 1920's progressively developed towards three-dimensional creations either with curved shapes or a more architectural design.

In 1933 Louis Cartier gave Jeanne Toussaint the highly important assignment of guiding the house designers and maintaining Cartier's standards of perfection, appointing her head of Fine Jewellery. During the dark years of the Depression and World War II, she deployed her rich imagination and intelligence, developing a new naturalistic style at Cartier. Both animals and flowers joined this marvelous, imaginary world. The supremacy of white metal was weakening as yellow gold made a return to the spotlight. New and daring colour combinations were proposed: citrine with amethyst (see BT 105 A39) or amethyst and aqua-marine (see CL 132 A37).

Delicate diamond-set flowers were designed, two becoming the favorite jewels of Queen Elizabeth of England and her sister HRH the Princess Margaret (see CL 296 A38). Coral remains a constant material in Cartier's creations, involving articulated settings that imitate nature as closely as possible. These techniques were put to test in 1968 when the untamable femme fatal, Mexican actress Maria Felix, challenged Cartier to devise a Snake necklace (see NE 10 A68). For her the Parisian workshop created amazing spectacular feat of art, composed of 2473 diamonds, 57 cm long, weighing 600 grams; it almost seemed to move of its own volition. Several years later she would spur Cartier to make one of its most fantastic pieces, composed of two crocodiles set with thousands of diamonds and emeralds, wearable as either brooches or as a necklace crossing their tails and heads (see NE 43 A75). The two reptiles are of such flexibility and realism that they seem alive when held in one's hand, a unique hallmark in the legacy of jewellery.

Pascale Lepeu
Curator of the Cartier Collection

62

JA 11 C1899

腰带扣

约 1899 年
卡地亚巴黎
银镀金
8.30 × 13.0 厘米

62

JA 11 C1899

Belt buckle

Cartier Paris, c. 1899
Silver-gilt
8.30 × 13.0 cm

发饰

1902 年
卡地亚巴黎
铂金
旧式和玫瑰式切割钻石
种子式镶嵌
中央高度　7.0 厘米

售予威廉·菲尔德夫人。

里拉·范德比尔特·菲尔德（原姓斯洛恩，1878～1934年）是艾米莉·托恩·范德比尔特和威廉·道格拉斯·斯洛恩的女儿，玛尔伯勒公爵夫人康斯薇洛·范德比尔特（后来的杰奎.巴尔桑夫人）的表妹。她嫁给了威廉·布拉德赫斯特·奥斯古德·菲尔德（来源：苏富比纽约，1999年10月20日销售目录）。

63

HO 25 A02

Hair ornament

Cartier Paris, 1902
Platinum
Old- and rose-cut diamonds
Millegrain setting
Height at centre 7.0 cm

Sold to Mrs. William Field

Lila Vanderbilt Field (born Sloane, 1878-1934) was the daughter of Emily Thorne Vanderbilt and William Douglas Sloane. The cousin of Consuelo Vanderbilt, Duchess of Marlborough (later Mme. Jacques Balsan), she married William Bradhurst Osgood Field (source: Sotheby's New York, sale catalogue of October 20, 1999).

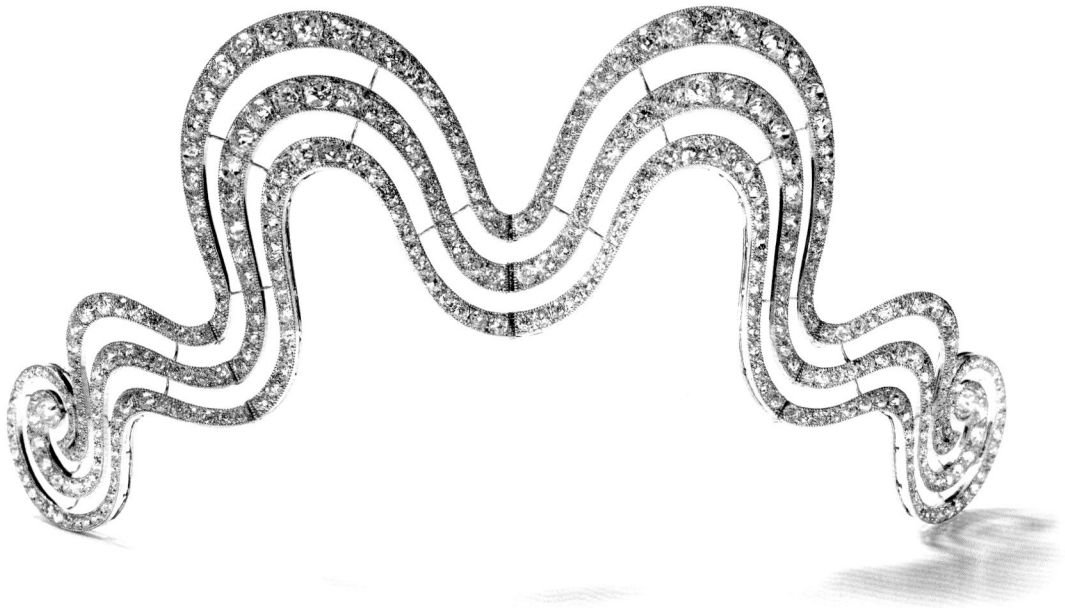

阳伞手柄

1902 年
卡地亚巴黎
铂金, 黄金, 玫瑰金
半透明灰蓝色、绿色、粉色、白色珐琅
明亮式和玫瑰式切割钻石
6.44 × 2.50 厘米

8个伞钮, 用作阳伞伞骨的顶帽。

64

AL 121 A02

Parasol handle

Cartier Paris, 1902
Platinum, yellow gold, pink gold
Translucent gray-blue enamel over
guilloché ground, green, pinkish and white
enamel
Brilliant- and rose-cut diamonds
6.44 × 2.50 cm

The eight parasol studs are used as
tips for the parasol's frame.

65

CL 288 A04

胸针

1904 年
卡地亚巴黎
铂金，金
欧洲旧式和玫瑰式切割钻石
切面红宝石
4.22 × 1.84 厘米

这款胸针与CL 300 A08一样，均标志着卡地亚现代风格的诞生，而花环风格则是这一阶段的顶峰。

65

CL 288 A04

Brooch

Cartier Paris, 1904
Platinum, gold
Old European- and rose-cut diamonds
Facetted rubies
4.22 × 1.84 cm

This brooch, as brooch CL 300 A08, both indicate the birth of Cartiers' Modern style, while the Garland style was at its apogee.

66

OV 10 A04

小型相机形相框

1904 年
卡地亚巴黎
黄金，白金，玫瑰金
凸圆形红宝石
剖面珍珠，绿色、白色珐琅
木
10.5 × 2.96 × 3.90 厘米

66

OV 10 A04

Photo frame in the form of a miniature camera

Cartier Paris, 1904
Yellow gold, white gold, pink gold
Ruby cabochons
Split pearls, green and white enamel
Wood
10.5 × 2.96 × 3.90 cm

企鹅（两只）

约 1904 年
卡地亚巴黎
雕花玛瑙
金
凸圆形红宝石（眼睛）
8.05 × 6.1 × 2.8 厘米

67

AN 15 C04

Sculpture of two penguins

Cartier Paris, c.1904
Carved agate
Gold
Ruby cabochons (eyes)
8.05 × 6.1 × 2.8 cm

68

AN 12 C04

兔

约 1904 年
卡地亚
雕花紫水晶
银
玫瑰式切割钻石（项圈）
凸圆形红宝石（眼睛）
3.6 × 2.1 × 3.9 厘米

68

AN 12 C04

Seated rabbit sculpture

Cartier, c. 1904
Carved amethyst
Silver
Rose-cut diamonds (collar)
Ruby cabochons (eyes)
3.6 × 2.1 × 3.9 cm

69

AN 11 C04

牛头犬

约 1904 年
卡地亚
雕花烟色石英石
金
蓝宝石，一颗巴洛克珍珠
橄榄石（眼睛）
6.59 × 3.63 × 5.15 厘米

69

AN 11 C04

Seated bulldog sculpture

Cartier, c. 1904
Carved smoky quartz
Gold
Sapphires, one Baroque pearl
Olivine (eyes)
6.59 × 3.63 × 5.15 cm

猪

约 1905 年
卡地亚
雕花玫瑰石
金
玫瑰式切割钻石（眼睛）
4.07 × 3.07 × 7.08 厘米

70

AN 16 C05

Seated pig sculpture

Cartier, c. 1905
Carved rhodonite
Gold
Rose-cut diamonds (eyes)
4.07 × 3.07 × 7.08 cm

71

HO 21 A05

发梳

1905 年
卡地亚
铂金，银，金
玳瑁
玫瑰式切割钻石
半透明蓝色、白色珐琅
种子式镶嵌
6.30 × 9.70 厘米

71

HO 21 A05

Hair comb

Cartier, 1905
Platinum, silver, gold
Tortoiseshell
Rose-cut diamonds
Translucent blue enamel over guilloché
ground, white enamel
Millegrain setting
6.30 × 9.70 cm

72

WN 16 A05

球形表吊坠项链

1905 年
卡地亚巴黎
铂金, 金
半透明粉红色、白色珐琅
珍珠, 玫瑰式切割钻石
项链　50.0 厘米
球形表直径　2.3 厘米

圆形机芯, 镀金, 15颗宝石轴承, 瑞士杠杆式擒纵结构, 双金属平衡摆轮, 扁平摆轮游丝。旋转外圈即可上链, 按压钻石环上的按钮同时旋转外圈, 即可设定时间。

72

WN 16 A05

Ball-shaped watch on
necklace

Cartier Paris, 1905
Platinum, gold
Translucent pink enamel over guilloché
ground, white enamel
Pearls, rose-cut diamonds
Necklace 50.0 cm; diameter of the ball-
shaped watch 2.3 cm

Round movement, gold-plated, 15 jewels, Swiss lever escapement, bi-metallic balance, flat balance spring. Winding is done by turning the bezel; time is set by turning the bezel while holding a button set in the ring of diamonds.

手链

1906 年
卡地亚巴黎
铂金，金
玫瑰式切割钻石
天然珍珠
半透明粉红色珐琅
种子式镶嵌
长度　17.0 厘米

搭扣上的玫瑰式切割钻石图案代表古斯拉夫语中的字母"J"。

73

BT 53 A06

Bracelet

Cartier Paris, 1906
Platinum, gold
Rose-cut diamonds
Natural pearls
Translucent pink enamel over guilloché
ground
Millegrain setting
Length 17.0 cm

The rose-cut diamond motif on the clasp represents the letter "J" in the Cyrillic alphabet.

74

CL 253 A06

胸针

1906 年
卡地亚巴黎
铂金，金
圆形旧式切割钻石
一颗圆形切面蓝宝石
切面和圆柱形切割蓝宝石
种子式镶嵌和轨道式镶嵌
3.03 × 6.0 厘米

74

CL 253 A06

Brooch

Cartier Paris, 1906
Platinum, gold
Round old-cut diamonds
One round faceted sapphire
Faceted and calibré-cut sapphires
Millegrain setting and channel-setting
3.03 × 6.0 cm

长柄眼镜

1906 年
卡地亚巴黎
金，铂金
玫瑰式切割钻石
半透明蓝色、象牙色珐琅
镜片直径　3.33 厘米
手柄（不含吊环）　7.03 厘米

带铰接框，不用时可折叠。

75

OI 06 A06

Lorgnette

Cartier Paris, 1906
Gold, platinum
Rose-cut diamonds
Translucent blue enamel over guilloché
ground, ivory-coloured enamel
Diameter of the glass 3.33 cm; the handle
(without the 2 suspension rings) 7.03 cm

The hinged frames enable them to be
folded away when not in use.

PE 05 A06

圣·安东尼·德·巴杜吊坠

1906 年
卡地亚巴黎
金，银，铂金
半透明蓝色珐琅
玫瑰式切割蓝宝石，玫瑰式切割钻石
3.62 × 1.37 厘米

800年来，圣·安东尼·德·巴杜就一直是最受欢迎的基督教圣人之一。在全球几乎所有教堂都能看到他怀抱婴儿的雕塑。信徒常祈请他帮助寻找失物。

76

PE 05 A06

Pendant *Saint Antoine de Padoue*

Cartier Paris, 1906
Gold, silver, platinum
Translucent blue enamel over guilloché ground
Rose-cut sapphires, rose-cut diamonds
3.62 × 1.37 cm

Since 800 years, Saint Antoine de Padoue (Saint Anthony of Padua), is one of the most popular Saints of the Christendom. A statue of him holding a child is displayed in almost all the churches throughout the world. The believers invoke him to find their lost objects.

猫头鹰晚装手袋

1906 年
卡地亚巴黎
金
玫瑰式切割钻石
四颗凸圆形祖母绿
18.00 × 17.05 × 1.05 厘米

Hibou evening bag

Cartier Paris, 1906
Gold
Rose-cut diamonds
Four emerald cabochons
18.00 × 17.05 × 1.05 cm

78

PB 17 A06

粉盒

1906 年
卡地亚巴黎
金，铂金
玫瑰式切割钻石
半透明粉红色珐琅
白色、蓝色、绿色珐琅
5.15 × 1.72 厘米

78

PB 17 A06

Powder box

Cartier Paris, 1906
Gold, platinum
Rose-cut diamonds
Translucent pink enamel over guilloché
ground
White, blue and green enamel
5.15 × 1.72 cm

79

SA 10 A06

烟灰缸

1906 年
卡地亚巴黎
银镀金
角闪岩（黝帘石）
白色、红色、黑色珐琅
9.02 × 6.75 × 2.95 厘米

79

SA 10 A06

Ashtray

Cartier Paris, 1906
Silver-gilt
Amphibolite (zoisite)
White, red and black enamel
9.02 × 6.75 × 2.95 cm

80

CDS 61 A06

支架式座钟

1906 年
卡地亚巴黎
金，银，银镀金，镀金属
半透明绿色、白色珐琅
10.65 × 12.15 厘米

此钟为8日储存圆形机芯，镀金，
瑞士杠杆式擒纵结构，双金属平衡摆
轮，宝玑摆轮游丝。

80

CDS 61 A06

Desk clock with Strut

Cartier Paris, 1906
Gold, silver, silver-gilt, gilded metal
Translucent green enamel over guilloché
ground, white enamel
10.65 × 12.15 cm

Round 8-day movement, gold-plated,
Swiss lever escapement, bimetallic
balance, Breguet balance spring.

81

IO 69 A06

卷尺盒

1906 年
卡地亚巴黎
金
半透明绿色、白色珐琅
玫瑰式切割钻石
直径　3.65 厘米

内有卷尺。

81

IO 69 A06

Tape measure case

Cartier Paris, 1906
Gold
Translucent green enamel over guilloché
ground, white enamel
Rose-cut diamonds
Diameter 3.65 cm

The interior is containing a tape
measure.

82

AN 10 A06

栖架上的猫头鹰

1906 年
卡地亚巴黎
雕花玛瑙（猫头鹰）
金，铂金，银镀金
玫瑰式切割钻石，凸圆形绿宝石，凸圆形蓝宝石
象牙（栖架）
绿色珐琅
7.08 × 3.09 厘米

售予罗斯柴尔德夫人。

82

AN 10 A06

Sculpture of an owl on a perch

Cartier Paris, 1906
Carved agate (owl)
Gold, platinum, silver-gilt
Rose-cut diamonds, emerald cabochons,
sapphire cabochons
Ivory (perch)
Green enamel
7.08 × 3.09 cm

Sold to Lady Rothschild

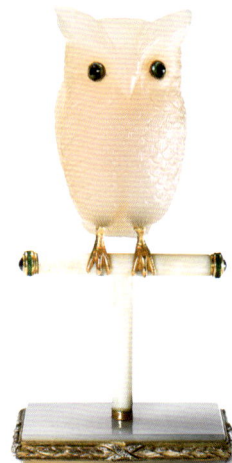

83

AN 06 C06

小鸡

约 1906 年
卡地亚
雕花玛瑙（小鸡），白色玉髓（蛋壳）
金
玫瑰式切割钻石（眼睛）
5.00 × 4.60 × 3.80 厘米

83

AN 06 C06

Chick sculpture

Cartier, c. 1906
Carved grey agate (chick), white chalcedony
(eggshell)
Gold
Rose-cut diamonds (eyes)
5.00 × 4.60 × 3.80 cm

84

BT 03 C07

手链

约 1907 年
卡地亚巴黎
金，铂金
圆形旧式切割钻石
切面和圆柱形切割红宝石
天然珍珠
长度　18.50 厘米

84

BT 03 C07

Line bracelet

Cartier Paris, c. 1907
Gold, platinum
Round old-cut diamonds
Faceted and calibré-cut rubies
Natural pearls
Length 18.50 cm

85

CL 23 A07

日本结胸针

1907 年
卡地亚巴黎
铂金，金
玫瑰式切割钻石
圆形切面和圆柱形切割红宝石
4.60 × 2.40 厘米

19世纪末期日本风格席卷欧洲。
这枚胸针体现了这一影响。

85

CL 23 A07

Japanese knot brooch

Cartier Paris, 1907
Platinum, gold
Rose-cut diamonds
Round faceted and calibré-cut rubies
4.60 × 2.40 cm

This brooch represents a highly original expression of the Japanese influence that swept through Europe from the late nineteenth century.

86

AL 119 A07

网状手袋（网眼钱包）

1907 年
卡地亚巴黎
金，铂金
凸圆形蓝宝石
绿色、白色珐琅
21.0 × 5.0 厘米

86

AL 119 A07

Filoche purse (mesh purse)

Cartier Paris, 1907
Gold, platinum
Sapphire cabochons
Green and white enamel
21.0 × 5.0 cm

CCI 04 A07

环形表盘座钟

1907 年
卡地亚巴黎
金，铂金，银，银镀金
半透明钢蓝色，半透明绿色，白色珐琅
玫瑰式切割钻石
8.1 × 6.0 厘米

售予安娜·古尔德。

此钟为8日储存圆形机芯，镀金，15颗宝石轴承，瑞士杠杆式擒纵结构，双金属平衡摆轮，宝玑摆轮游丝。

安娜·古尔德为美国铁路大亨之女，她亦是"美好年代"时期社交名流波尼·德·卡斯特兰的前妻。

87
CCI 04 A07

Desk clock with annular dial

Cartier Paris, 1907
Gold, platinum, silver, silver-gilt
Steel-blue translucent and translucent
green enamel over guilloché ground, white
enamel
Rose-cut diamonds
8.1 × 6.0 cm

Round 8-day movement, gold-plated, 15 jewels, Swiss lever escapement, bimetallic balance, Breguet balance spring.

Sold to Anna Gould, daughter of the American railroad magnate and ex-wife of the Belle époque socialite, Boni de Castellane.

路易十六屏风式时钟

1907 年
卡地亚巴黎
黄金, 玫瑰金, 银, 银镀金
半透明乳白色、蓝色、白色珐琅
凸圆形蓝宝石
9.55 × 3.70 × 4.55 厘米

此钟为8日储存圆形机芯, 镀金,
15颗宝石轴承, 瑞士杠杆式擒纵结构,
双金属平衡摆轮, 宝玑摆轮游丝。

88

CDS 50 A07

Small Louis XVI screen
clock

Cartier Paris, 1907
Yellow gold, pink gold, silver, silver-gilt
Translucent opaline-blue enamel over
guilloché ground, white enamel
Sapphire cabochon
9.55 × 3.70 × 4.55 cm

Round 8-day movement, gold-plated,
15 jewels, Swiss lever escapement,
bimetallic balance, Breguet balance
spring.

89

IO 37 A07

针枕

1907 年
卡地亚巴黎
银，金
半透明紫色、绿色、白色珐琅
天鹅绒
3.2 × 5.9 厘米

89

IO 37 A07

Pin cushion

Cartier Paris, 1907
Silver, gold
Translucent purple and green enamel over
guilloché ground, white enamel
Velvet
3.2 × 5.9 cm

须梳

约 1907 年
卡地亚巴黎
金
玳瑁
蓝色和白色条纹珐琅装饰
7.00 × 1.05 厘米

90

AG 83 C07

Moustache comb

Cartier Paris, c. 1907
Gold
Tortoiseshell
Striped blue-and-white enamel decoration
(pékin)
7.00 × 1.05 cm

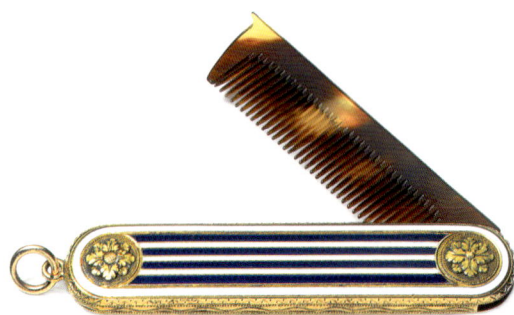

香熏

1907 年
卡地亚巴黎
银镀金
砂金石
凸圆形蓝宝石
半透明蓝色、白色珐琅
13.1 × 5.63 厘米

91

FK 16 A07

Perfume burner

Cartier Paris, 1907
Silver-gilt
Aventurine
Sapphire cabochons
Translucent blue enamel over guilloché
ground, white enamel
13.1 × 5.63 cm

花瓶（一对）

1907 年
卡地亚巴黎
银镀金，金
玫瑰色石英
半透明蓝色珐琅
象牙
10.9 × 3.86 × 3.86 厘米

Pair of vases

Cartier Paris, 1907
Silver-gilt, gold
Rose quartz
Translucent blue enamel over guilloché
ground
Ivory
10.9 × 3.86 × 3.86 cm

93

AN 08 C07

牛头犬

约 1907 年

卡地亚

雕花石英石

金

玫瑰式切割钻石（眼睛）

6.6 × 5.5 × 4.0 厘米

93

AN 08 C07

Seated bulldog sculpture

Cartier, c. 1907

Carved smoky quartz

Gold

Rose-cut diamonds (eyes)

6.6 × 5.5 × 4.0 cm

94

HO 03 A08

卡柯史尼克冠冕

1908 年
卡地亚巴黎
铂金
15 颗梨形钻石（重约 19 克拉）
圆形旧式切割钻石
天然珍珠
"铃兰"镶嵌梨形垂坠钻石
中央高度　5.0 厘米

此款冠冕中的垂坠钻石展示了珠宝史上最早期的"铃兰"镶嵌技法。以纤巧的铂金，打造为铃兰（山谷百合）的形状，环绕成为铃兰形状；中间镶嵌一颗明亮式切割大钻石，四周紧密簇拥数颗钻石，看起来就如单颗巨型钻石。这种镶嵌法在1910年左右达臻完美。

94

HO 03 A08

Kokoshnik tiara

Cartier Paris, 1908
Platinum
Fifteen pear-shaped diamonds, weighing
approximately 19 carats in total
Round old-cut diamonds
Natural pearls
"Muguet" setting for the swinging pear-
shaped diamonds
Height at centre 5.0 cm

The hanging diamonds represent one of the earliest examples of the technique known as a "muguet" setting. A delicate rim of platinum in the shape of a lily-of-the-valley (muguet) holds a large brilliant surrounded by several tightly packed diamonds that create the impression of a single stone. This type of setting was perfected around 1910.

95

CL 300 A08

胸针

1908 年
卡地亚巴黎
铂金
单面切割钻石
圆柱形切割蓝宝石
天然珍珠
直径 5.33 厘米

（参见CL 288 A04）

95

CL 300 A08

Brooch

Cartier Paris, 1908
Platinum
Single-cut diamonds
Calibré-cut sapphires
Natural pearls
Diameter 5.33 cm

(see CL 288 A04)

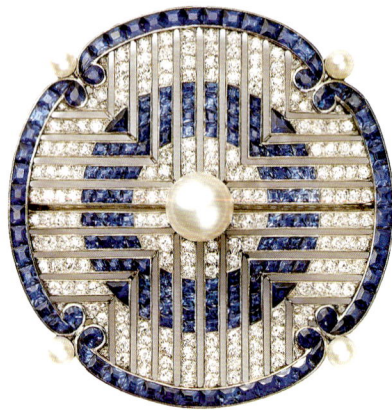

96

WP 17 A08

表吊坠

1908 年
卡地亚巴黎
铂金，金
半透明铁灰色珐琅
一颗玫瑰式切割红宝石，一颗凸圆形红宝石，圆
柱形切割红宝石
玫瑰式切割钻石
3.9 × 3.1 × 0.4 厘米

此表为圆形机芯，镀金，18颗宝石
轴承，瑞士杠杆式擒纵结构，双金属平
衡摆轮，扁平摆轮游丝。

96

WP 17 A08

Pendant watch

Cartier Paris, 1908
Platinum, gold
Translucent iron-gray enamel over guilloché
ground
One rose-cut ruby, one ruby cabochon,
calibré-cut rubies
Rose-cut diamonds
3.9 × 3.1 × 0.4 cm

Round movement, gold-plated, 18
jewels, Swiss lever escapement, bimetallic balance, flat balance spring.

97

HO 15 A08

束发带

1908 年
卡地亚巴黎
金
珍珠
玳瑁
半透明蓝色珐琅
直径　10.7 厘米

97

HO 15 A08

Bandeau

Cartier Paris, 1908
Gold
Pearls
Tortoiseshell combs
Translucent blue enamel over guilloché
ground
Diameter 10.7 cm

98

JA 23 A08

帽针

1908 年
卡地亚巴黎
铂金，金
半透明蓝色、白色珐琅
圆形旧式和玫瑰式切割钻石
12.50 × 3.0 厘米

98

JA 23 A08

Hat pin

Cartier Paris, 1908
Platinum, gold
Translucent blue enamel over guilloché
ground, white enamel
Round old- and rose-cut diamonds
12.50 × 3.0 cm

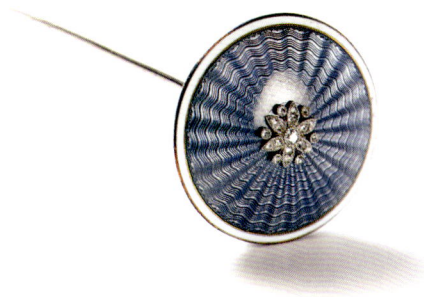

99

CDB 20 A08

方尖碑时钟

1908 年
卡地亚巴黎
银
电气石（方尖碑）
半透明乳白色珐琅
绿色、白色珐琅
29.2 × 7.5 × 7.5 厘米

此钟为8日储存圆形机芯，镀金，12颗宝石轴承，瑞士杠杆式擒纵结构，双金属平衡摆轮，扁平摆轮游丝。上链和时间调校轴位于背后，旋转机动雕花装饰和半透明绿色珐琅银盘即可。

这款时钟最初高度为39厘米，1910年被缩短。最初采用的是铂金镶玫瑰式切割钻石指针，后期经过更换。

99

CDB 20 A08

Obelisk clock

Cartier Paris, 1908
Silver
Fluorine (obelisk)
Translucent opaline enamel over guilloché ground
Green and white enamel
29.2 × 7.5 × 7.5 cm

Round 8-day movement, gold-plated, 12 jewels, Swiss lever escapement, bimetallic balance, flat balance spring. Arbor for setting hands and winding movement on back, accessed by pivoting a silver disk with engine-turned decoration and translucent green enameling.

Originally 39 centimeters high, this clock was shortened in 1910. The original hands of platinum and rose-cut diamonds were replaced at a later date.

100

CCI 05 A08

环形表盘时钟

1908 年
卡地亚巴黎
铂金，银，银镀金
玛瑙（花瓶瓶身）
玫瑰式切割钻石，凸圆形蓝宝石
乳白色、灰蓝色、半透明绿色、白色珐琅
14.8 × 5.4 厘米

此钟为8日储存圆形机芯，镀金，瑞士杠杆式擒纵结构，双金属平衡摆轮。时间调校和上链装置位于右侧把手上。

瓶身为路易十六风格。

100

CCI 05 A08

Clock with inner annular dial

Cartier Paris, 1908
Platinum, silver, silver-gilt
Agate (vase)
Rose-cut diamonds, sapphire cabochon
Opaline gray-blue and green translucent
enamel over guilloché ground, white
enamel
14.8 × 5.4 cm

Round 8-day movement, gold-plated, Swiss lever escapement, bimetallic balance. Hand-setting and winding mechanism on handle at right.

The vase is of Louis XVI style.

IOI

DI 24 A08

墨水瓶时钟

1908 年
卡地亚巴黎
金、银、银镀金
凸圆形蓝宝石, 玫瑰式切割钻石
半透明灰色、紫色、浅紫色、绿色、白色珐琅
18.50 × 7.00 × 8.30 厘米

此钟为8日储存长方形机芯, 镀金, 瑞士杠杆式擒纵结构, 双金属平衡摆轮, 扁平摆轮游丝。设有上链和时间调校轴。

IOI

DI 24 A08

Clock on inkstand

Cartier Paris, 1908
Gold, silver, silver-gilt
Sapphire cabochons, rose-cut diamonds
Translucent gray, mauve, pale mauve, and green enamel over guilloché ground, white enamel
18.50 × 7.00 × 8.30 cm

Rectangular 8-day movement, gold-plated, Swiss lever escapement, bi-metallic balance, flat balance spring. Arbor for winding movement and setting hands.

FL 03 C08

铃兰案头装饰

约 1908 年
卡地亚巴黎
金
雕花玛瑙（壶），玫瑰色石英（基座）和砂金石（叶）
月长石
珍珠，珐琅涂层人造宝石（花朵）
象牙（架子）
黄铜，玻璃和木（外罩）
18 × 10.9 × 10.9 厘米

FL 03 C08

Lilly of the valley table

ornament

Cartier Paris, c. 1908
Gold
Carved agate (pot), rose quartz (base) and
aventurine (leaves)
Moonstone
Oriental pearls, enamelled paste (flowers)
Ivory (stand)
Brass, glass and wood (showcase)
18 × 10.9 × 10.9 cm

103

WPO 10 A09

项链式怀表

1909 年
卡地亚巴黎
铂金，金
半透明紫色、白色珐琅
珍珠，玫瑰式切割钻石
直径　4.70 厘米

此表为积家142型圆形机芯，镀金，18颗宝石轴承，瑞士杠杆式擒纵结构，双金属平衡摆轮，扁平摆轮游丝。

103

WPO 10 A09

Pocket watch with necklace

Cartier Paris, 1909
Platinum, gold
Translucent purple enamel over guilloché ground, white enamel
Pearls, rose-cut diamonds
Diameter 4.70 cm

Round LeCoultre caliber 142 movement, gold-plated, 18 jewels, Swiss lever escapement, bimetallic balance, flat balance spring.

104

WN 12 A09

项链式挂表

1909 年
卡地亚巴黎
铂金，金
玫瑰式切割钻石
半透明蓝色珐琅
珍珠
直径　2.94 厘米

此表为积家10HPVM型圆形机芯，镀金，18颗宝石轴承，瑞士杠杆式擒纵结构，双金属平衡摆轮，扁平摆轮游丝。

104

WN 12 A09

Watch on necklace

Cartier Paris, 1909
Platinum, gold
Rose-cut diamonds
Translucent blue enamel over guilloché ground
Pearls
Diameter 2.94cm

Round LeCoultre caliber 10HPVM movement, gold-plated, 18 jewels, Swiss lever escapement, bimetallic balance, flat balance spring.

105

CL 102 A09

胸针

1909 年
卡地亚巴黎
铂金，金
圆形旧式切割钻石
方形切面和圆柱形切割红宝石
种子式镶嵌
4.11 × 2.6 厘米

105

CL 102 A09

Brooch

Cartier Paris, 1909
Platinum, gold
Round old-cut diamonds
Square faceted and calibré-cut rubies
Millegrain setting
4.11 × 2.6 cm

钢笔

1909 年
卡地亚巴黎
金
黑色和白色条纹珐琅装饰
长度　7.74 厘米

106

WI 04 A09

Fountain pen

Cartier Paris, 1909
Gold
Striped white-and-black enamel decoration
(*pékin*)
Length 7.74 cm

107
CL 221 A10

蝴蝶结胸针

1910 年
卡地亚巴黎
铂金
圆形旧式切割钻石
圆柱形切割黑玛瑙
种子式镶嵌
5.50 × 2.0 厘米

107
CL 221 A10

Bow brooch

Cartier Paris, 1910
Platinum
Round old-cut diamonds
Calibré-cut onyxes
Millegrain setting
5.50 × 2.0 cm

CL 202 A10

胸针

1910 年
卡地亚巴黎，特别订制
铂金
圆形旧式切割钻石
切面和圆柱形切割蓝宝石
种子式镶嵌
直径　4.15 厘米

售予康内留斯·范德比尔特夫人。

最早是一枚帽针，镶嵌明亮式切割钻石组成的波浪图案，为罗斯柴尔德家族订制，后经卡地亚改款为胸针。

康内留斯·范德比尔特三世（1873～1942年）是美国铁路大亨康内留斯·范德比尔特之曾孙。

108

CL 202 A10

Brooch

Cartier Paris, special order, 1910
Platinum
Round old-cut diamonds
Faceted and calibré-cut sapphires
Millegrain setting
Diameter 4.15 cm

Originally a hat pin with a wave motif of brilliants, made for the Rothschild family, this was later transformed by Cartier into a brooch.

Sold to Mrs. Cornelius Vanderbilt

Mr. Cornelius Vanderbilt III (1873-1942) was the great-grandson of "Commodore" Cornelius Vanderbilt, the U.S. railroad magnate.

NE 16 A10

项链

1910 年
卡地亚巴黎
铂金
单面切割、圆形旧式切割和玫瑰式切割钻石
切面和圆柱形切割蓝宝石
天然珍珠
种子式镶嵌
总长度　44.3 厘米
吊坠长度　13.0 厘米

来源于玛丽·司各特·汤森特夫人
和唐纳德·麦克罗伊夫人 (参见CL 134
A06, NE 39 A06)。

珍珠由小到大，均匀排列，为这款
吊坠赋予了绝佳动感。

109

NE 16 A10

Necklace

Cartier Paris, 1910
Platinum
Single-, round old- and rose-cut diamonds
Faceted and calibré-cut sapphires
Natural pearls
Millegrain setting
Total length 44.3 cm; length of the pendant
13.0 cm

The carefully graduated size of the
pearls lends great dynamism to this
pendant.

Provenance: Mary Scott Townsend
and Mrs. Donald McElroy

(see CL 134 A06, NE 39 A06)

CC 96 A10

烟盒

1910 年
卡地亚巴黎
金
半透明粉红色、白色珐琅
一颗凸圆形蓝宝石
11.05 × 3.82 厘米

110

CC 96 A10

Cigarette case

Cartier Paris, 1910
Gold
Translucent pink enamel over guilloché
ground, white enamel
One sapphire cabochon
11.05 × 3.82 cm

III

PF 13 C10

相框

约 1910 年
卡地亚巴黎
银镀金
玛瑙饰板
彩色宝石
17.0 × 10.80 厘米

宝石象征"十年",或是青年女子
的名字。

III

PF 13 C10

Photo frame

Cartier Paris, c. 1910
Silver-gilt
Agate plaque
Varicoloured stones
17.0 × 10.80 cm

The stones symbolizing "10 years" or
may be the young girl's first name.

112

AN 03 C10

栖架上的玛瑙长尾鹦鹉（一对）

约 1910 年
卡地亚巴黎
雕花玛瑙
玫瑰色石英（基座）
玫瑰式切割钻石（眼睛）
金
象牙（栖架）
18.5 × 8.00 厘米

112

AN 03 C10

Pair of hardstone
budgerigars on a perch

Cartier Paris, c. 1910
Carved hardstone
Rose quartz (base)
Rose-cut diamonds (eyes)
Gold
Ivory (perch)
18.5 × 8.00 cm

113

AN 01 C11

猪

约 1911 年
卡地亚巴黎
雕花玫瑰色石英
蓝宝石
6.73 × 3.95 × 2.90 厘米

113

AN 01 C11

Pig sculpture

Cartier Paris, c. 1911
Carved rose quartz
Sapphires
6.73 × 3.95 × 2.90 cm

CDS 65 A11

时钟

1911 年
卡地亚巴黎
银，银镀金，铂金
玛瑙（基座）
半透明粉红色、白色珐琅
玫瑰式切割钻石
4.98 × 3.65 × 2.94 厘米

此钟为8日储存长方形机芯，镀金，瑞士杠杆式擒纵结构，双金属平衡摆轮，宝玑摆轮游丝。设有上链和时间调校轴。

II4

CDS 65 A11

Mignonnette clock

Cartier Paris, 1911
Silver, silver-gilt, platinum
Agate (base)
Translucent pink enamel over guilloché ground, white enamel
Rose-cut diamonds
4.98 × 3.65 × 2.94 cm

Rectangular 8-day movement, gold-plated, Swiss lever escapement, bimetallic balance, Breguet balance spring. Arbor for winding movement and setting hands.

115

WP 04 A12

胸针表

1912 年
卡地亚巴黎
铂金，金
金色和白色细条纹珐琅装饰
玫瑰式切割钻石，圆柱形切割蓝宝石
白色珐琅
表直径　2.9 厘米

此表为圆形积家119型机芯，镀金，8个调校项目，18颗宝石轴承，瑞士杠杆式擒纵结构，双金属平衡摆轮，扁平摆轮游丝。

115

WP 04 A12

Watch-brooch

Cartier Paris, 1912
Platinum, gold
Striped gold-and-white enamel decoration
(*pékin*)
Rose-cut diamonds, calibre-art sapphires
White enamel
Diameter of the watch 2.9 cm

Round LeCoultre caliber 119 movement, gold-plated, 8adjustments, 18jewels, Swiss lever escapement, bi-metallic balance, flat balamce spring.

116

CL 249 A12

菱形胸针

1912 年
卡地亚巴黎
铂金
圆形旧式切割钻石
长方形、三角形、正方形切面蓝宝石
天然珍珠
4.6 × 3.8 厘米

售予罗斯柴尔德家族成员（请参见BT 108 A13, BT 104 A53）。

116

CL 249 A12

Lozenge brooch

Cartier Paris, 1912
Platinum
Round old-cut diamonds
Rectangular, triangular and square faceted
sapphires
Natural pearls
4.6 × 3.8 cm

Sold to a member of the Rothschild
family
(see BT 108 A13, BT 104 A53)

117

CL 197 A12

胸针吊坠

1912 年
卡地亚巴黎
铂金，金
圆形旧式和单面切割钻石
三颗凸圆形蓝宝石（其中一颗为菱形，重 28.70
克拉）
切面和圆柱形切割蓝宝石
7.35 × 3.96 厘米

117

CL 197 A12

Brooch-pendant

Cartier Paris, 1912
Platinum, gold
Round old- and single-cut diamonds
Three sapphire cabochons (one lozenge-
shaped, weighing 28.70 carats)
Faceted and calibré-cut sapphires
7.35 × 3.96 cm

118

CC 74 A12

烟盒

1912 年

卡地亚巴黎

玛瑙

铂金

玫瑰式切割钻石

圆柱形切割蓝宝石和凸圆形蓝宝石

9.30 × 5.05 × 1.40 厘米

正面饰有字母组合图案。

118

CC 74 A12

Cigarette case

Cartier Paris, 1912

Agate

Platinum

Rose-cut diamonds

Calibré-cut sapphires and sapphire
cabochons

9.30 × 5.05 × 1.40 cm

The front applied with a monogram.

119

AN 17 A12

牛头犬

1912 年
卡地亚
雕花玛瑙
金
珍珠
橄榄石（眼睛）
3.14 × 4.93 厘米

119

AN 17 A12

Standing bulldog sculpture

Cartier, 1912
Carved agate
Gold
Pearl
Olivine (eyes)
3.14 × 4.93 cm

120

BT 108 A13

手链

1913 年
卡地亚巴黎
铂金，金
玫瑰式切割钻石
凸圆形蓝宝石和面弧形蓝宝石
圆形切面黄玉
种子式镶嵌
长度　17 厘米

售予罗斯柴尔德家族成员（参见 CL 249 A12，BT 104 A53）。

这条手链是"握手"手链订单的一部分。该订单一共订制了10条一样的手链。

面弧形宝石是带有轻浅切面的凸圆形宝石，边缘经过抛光磨圆。

120

BT 108 A13

Bracelet

Cartier Paris, 1913
Platinum, gold
Rose-cut diamonds
Sapphire cabochons and buff-top sapphires
Round faceted topazes
Millegrain setting
Length 17 cm

This bracelet was part of an order of ten identical "handshake" bracelets.

Buff-top stones are lightly faceted cabochons with edges rounded by polishing.

Sold to a member of the Rothschild family

(see CL 249 A12, BT 104 A53)

121

NE 01 A13

埃及风格吊坠

1913 年
卡地亚巴黎
铂金
三角形、梨形、圆形旧式切割、单面切割和玫瑰
式切割钻石
圆柱形和花式切割黑玛瑙
种子式镶嵌
4.95 × 4.9 × 0.37 厘米

塔门或庙门图案,常被埃及金匠
用于吊坠之中,作为胸饰佩戴。

121

NE 01 A13

Egyptian-style pendant

Cartier Paris, 1913
Platinum
Triangular, pear-shaped, round old-, single-
and rose-cut diamonds
Calibré- and fancy-cut onyxes
Millegrain setting
4.95 × 4.9 × 0.37 cm

The motif of the pylon, or monumental temple gate, was used by Egyptian
goldsmiths for pendants worn as pec-
toral jewellery.

胸针

1913 年
卡地亚巴黎
铂金
圆形旧式切割钻石
一颗三角斜面形祖母绿(重 11.90 克拉)
凸圆形祖母绿和水滴形祖母绿
天然珍珠
黑玛瑙
9.7 × 3.6 厘米

这款胸针曾于1913年11月在卡
地亚纽约所举行的"卡地亚珠宝创
作——来自印度、波斯、阿拉伯、俄国
和中国的艺术"中展出。

122

CL 183 A13

Brooch

Cartier Paris, 1913
Platinum
Round old-cut diamonds
One triangular beveled 11.90 carat emerald
Emerald cabochons and drop-shaped
emeralds
Natural pearls
Onyx
9.7 × 3.6 cm

This brooch was exhibited in the *"Col-
lection of Jewels··· created by Messieurs
Cartier··· from the Hindoo, Persian,
Arab, Russian and Chinese Arts"* at
Cartier New York in November 1913.

123

CL 193 A13

胸针吊坠

1913 年
卡地亚巴黎
铂金
单面切割和玫瑰式切割钻石
凸圆形切割蓝宝石
一颗天然珍珠
凸圆形切割绿松石
九块相互链接的翡翠片
12.30 × 3.98 厘米

123

CL 193 A13

Brooch-pendant

Cartier Paris, 1913
Platinum
Single- and rose-cut diamonds
Sapphire cabochons
One natural pearl
Turquoise cabochons
Nine hinged jade plaques
12.30 × 3.98 cm

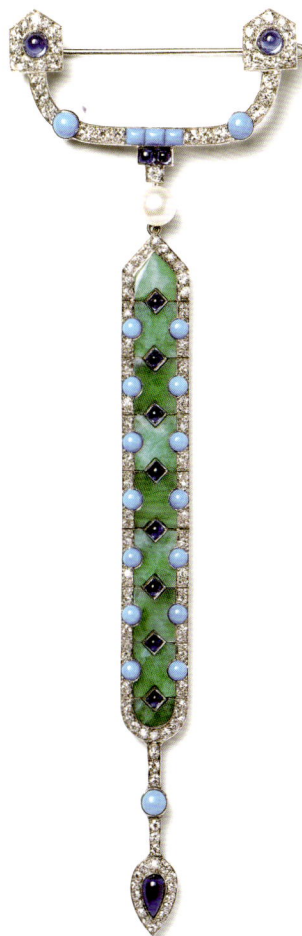

124

CS 08 A13

昼夜彗星时钟

1913 年
卡地亚巴黎
金, 铂金, 镀银金属
玫瑰式切割钻石
无色水晶
半透明长春花蓝色、白色珐琅
白色珐琅条纹
直径　9.5 厘米

此钟为8日储存圆形机芯, 镀金, 15颗宝石轴承, 瑞士杠杆式擒纵结构, 双金属平衡摆轮, 宝玑摆轮游丝。表盘中央每24小时旋转一次, 分钟指针与表盘后隐藏的指针顶端相连。

在这款时钟中, 整点时间以两个镶嵌玫瑰式切割钻石的铂金指示器指示。这两个指针与中央相连, 轮流显示, 太阳代表白天, 新月代表夜晚。

124

CS 08 A13

Day and night *comet* clock

Cartier Paris, 1913
Gold, platinum, silver-plated metal
Rose-cut diamonds
Rock crystal
Translucent periwinkle-blue and white enamel
Striped white enamel decoration (*pékin*)
Diameter 9.5 cm

Round 8-day movement, gold-plated, 15 jewels, Swiss lever escapement, bimetallic balance, Breguet balance spring. The center of the dial rotates once every 24 hours. The minute pointer is attached to the tip of a hand hidden beneath the dial.

In this clock the hours are indicated by 2 pointers of rose-cut diamonds set in platinum, attached to the centre and appearing in turn, a sun in daytime and a crescent moon at night.

125

HO 27 A14

冠冕

1914 年
卡地亚巴黎
铂金
圆形旧式切割钻石
15 颗天然珍珠
黑玛瑙镶嵌树形图案
黑色珐琅
中央高度 4.3 厘米

这款作品取材于俄国卡柯史尼克
冠冕的造型, 风格大胆前卫, 是装饰艺
术的绝佳体现。

125

HO 27 A14

Tiara

Cartier Paris, 1914
Platinum
Round old-cut diamonds
Fifteen natural pearls
Onyx stylized tree pattern
Black enamel
Height at centre 4.3 cm

Inspired by the shape of the Russian
kokoshnik tiaras, this audacious and
avant-garde jewel is a magnificent
example of the Art Deco stylization.

CL 255 A14

水果碗胸针

1914 年
卡地亚巴黎
金
凸圆形红宝石、切面圆柱形切割红宝石
凸圆形祖母绿
一颗凸圆形黑玛瑙
黑玛瑙（碗盘）
3.69 × 2.59 厘米

126

CL 255 A14

Fruit bowl brooch

Cartier Paris, 1914
Gold
Ruby cabochons, faceted calibré-cut rubies
Emerald cabochons
One onyx cabochon
Onyx (bowl)
3.69 × 2.59 cm

127

CC 61 A14

俄国烟盒

1914 年
卡地亚巴黎
玉，黑玛瑙（簧销）
金
黑色珐琅，黑色丝绳
11.00 × 6.50 厘米

售予W·B·利兹夫人。

烟盒内设两个隔层，一个放置香烟，一个放置火柴。

南希·利兹系美国锡业大亨W·B·利兹遗孀，也是卡地亚在第一次世界大战前最具魅力的客人之一。1910年，卡地亚以57万美元售予其一条珍珠镶钻项链登上了报纸头条。她的儿子在1921年迎娶了希腊齐妮亚公主。

127

CC 61 A14

Russian cigarette case

Cartier Paris, 1914
Nephrite, onyx (thumb piece)
Gold
Black enamel, black silk cord
11.00 × 6.50 cm

The interior fitted with two compartments, one for cigarettes, the other for matches.

Sold to Mrs. W.B Leeds

Widow of American tin tycoon W. B. Leeds, Nancy Leeds was one of Cartier's most fascinating clients up until World War I. Cartier hit the headlines in 1910 when the company sold her a pearl-and-diamond necklace for $570,000. Her son married Princess Xenia of Greece in 1921.

128

WB 33 A15

豹纹表胸针

1915 年
卡地亚巴黎
铂金
三颗梨形切割钻石、旧式切割钻石
黑玛瑙斑点和绳圈
双黑色软绳
长度　15.5 厘米

售予皮埃尔·卡地亚。

此表为积家111型长方形带截角机芯，模拟日内瓦波纹装饰，镀银，8个调校项目，19颗宝石轴承，瑞士杠杆式擒纵结构，双金属平衡摆轮，扁平摆轮游丝。

128

WB 33 A15

Panther-pattern watch-brooch

Cartier Paris, 1915
Platinum
Three pear-shaped diamonds, old-cut
diamonds
Onyx spots and slider
Double black cord
Length 15.5 cm

Rectangular LeCoultre caliber 111 movement with cut corners, fausses Côtes de Genève decoration, silver-plated, 8 adjustments, 19 jewels, Swiss lever escapement, bimetallic balance, flat balance spring.

Sold to Pierre Cartier

129

WCL 87 A15

Santos 腕表

1915 年
卡地亚巴黎
黄金, 玫瑰金
凸圆形蓝宝石
皮表带
3.49 × 2.47 厘米（表壳）

此表为积家126型圆形机芯, 日内
瓦波纹装饰, 镀银, 8个调校装置, 18颗
宝石轴承, 瑞士杠杆式擒纵结构, 双金
属平衡摆轮, 宝玑摆轮游丝。

129

WCL 87 A15

Santos wristwatch

Cartier Paris, 1915
Yellow gold, pink gold
Sapphire cabochon
Leather strap
3.49 × 2.47 cm (case)

Round LeCoultre caliber 126 movement, Côtes de Genève decoration,
silver-plated, 8 adjustments, 18 jewels,
Swiss lever escapement, bimetallic
balance, Breguet balance spring.

130

TC 06 C17

驾乘用品

约 1917 年
卡地亚纽约
金
黑色、白色珐琅
玻璃
皮革
木
20.0 × 10.6 × 8.6 厘米

前面嵌一个圆形时钟，内部设一面镜子，两个玻璃罐，两个椭圆形香精瓶，配以金及珐琅瓶盖，一个小皮夹和一本地址簿，一只金帽铅笔，一个针枕，一个皮革发梳盒。

130

TC 06 C17

Motorist necessaire

Cartier New York, c. 1917
Gold
Black and white enamel
Glass
Leather
Wood
20.0 × 10.6 × 8.6 cm

The front with an integrated clock of circular shape, the interior fitted with a mirror, two glass pots and two oblong perfume flasks with gold and enamel caps, a small leather purse and an address book, a gold capped pencil, a cushion pin, and a leather comb étui.

131

CL 285 A19

凯旋门胸针

1919 年
卡地亚巴黎
金，铂金
圆形旧式切割和玫瑰式切割钻石
凸圆形蓝宝石
花式切割切面红宝石和祖母绿
切面和圆柱形切割黄玉
黑玛瑙（阴影）
4.6 × 3.9 厘米

凸圆形蓝宝石代表1919年巴士底
日(7月14日，法国国庆日)沿香榭丽舍
大道游行、庆祝一战胜利的士兵所戴
的头盔。

131

CL 285 A19

Arc de Triomphe brooch

Cartier Paris, 1919
Gold, platinum
Round old- and rose-cut diamonds
Sapphire cabochons
Fancy-cut faceted rubies and emeralds
Faceted and calibré-cut topazes
Onyx (shade)
4.6 × 3.9 cm

The sapphire cabochons represent
the helmets of soldiers who paraded
down the Champs-Elysées to cel-
ebrate victory in World War One on
Bastille Day (July 14) 1919.

132

CL 257 A19

凯旋门旗帜胸针

1919 年
卡地亚巴黎
铂金，金
圆形旧式切割和玫瑰式切割钻石
切面圆柱形切割红宝石、蓝宝石和红锆石
花式切割切面祖母绿
一颗切面黑玛瑙
2.94 × 4.1 × 0.2 厘米

售予印度巴提亚拉土邦王公布品
德拉·塞恩勋爵。

132

CL 257 A19

Arc de Triomphe flag brooch

Cartier Paris, 1919
Platinum, gold
Round old- and rose-cut diamonds
Faceted calibré-cut rubies, sapphires and
jacinths
Fancy-cut faceted emeralds
One faceted onyx
2.94 × 4.1 × 0.2 cm

Sold to Sir Bhupindra Singh, Maharajah of Patiala

133

EG 01 A19

耳坠（一对）

1919 年
卡地亚巴黎为卡地亚纽约制作
铂金，金
圆形旧式切割和单面切割钻石
两颗黄色梨形蓝宝石（分别重 5.69 克拉 和 7.46
克拉）
天然珍珠
黑玛瑙（花冠形）
黑色珐琅
"铃兰"镶嵌悬垂钻石
8.20×1.50 厘米

133

EG 01 A19

Pair of ear-pendants

Cartier Paris for New York, 1919
Platinum, gold
Round old- and single-cut diamonds
Two pear-shaped yellow sapphires (5.69
and 7.46 carats)
Natural pearls
Onyx (corollae)
Black enamel
"Muguet"-setting for the hanging diamonds
8.20 × 1.50 cm

134

CL 181 A20

胸针

1920 年
卡地亚巴黎
铂金
圆形旧式切割和单面切割钻石
黑玛瑙
2.7 × 5.5 厘米

售予印度巴提亚拉土邦王公布品
德拉·塞恩勋爵。

134

CL 181 A20

Brooch

Cartier Paris, 1920
Platinum
Round old- and single-cut diamonds
Onyx
2.7 × 5.5 cm

Sold to Sir Bhupindra Singh, Maharajah of Patiala

135

PE 21 A20

双鸟饮泉吊坠

1920 年
卡地亚伦敦
铂金
一颗枕形钻石，圆形旧式切割、单面切割和玫瑰
式切割钻石
圆柱形和花式切割蓝宝石
无色水晶（水）
月长石（水瓮）
圆柱形切割黑玛瑙
5.8 × 5.4 × 0.8 厘米

这款吊坠设计于 1913 年，最早为搭配丝绳佩戴。在水瓮的两侧，各有一个钻石铺镶虎头，连接着一个黑玛瑙圆环。

135

PE 21 A20

Pendant with two birds drinking from fountain

Cartier London, 1920
Platinum
One cushion-shaped diamond, round old-,
single- and rose-cut diamonds
Calibré- and fancy-cut sapphires
Rock crystal (water)
Moonstone (vase)
Calibré-cut onyxes
5.8 × 5.4 × 0.8 cm

A late example of a model designed in 1913, this pendant was originally worn on a cord. On each side of the bowl, a diamond-paved tiger head holds a ring of onyx.

带烟管的化妆盒

1920 年
卡地亚巴黎
金，铂金
黑色和奶油色珐琅，金色和黑色条纹珐琅装饰
玫瑰式切割钻石
玳瑁（烟管）
11.7 × 4.4 × 3.0 厘米

136

VC 17 A20

Vanity case with cigarette holder

Cartier Paris, 1920
Gold, platinum
Black and cream-coloured enamel, striped
gold-and-black enamel decoration (pékin)
Rose-cut diamond
Tortoiseshell (cigarette holder)
11.7 × 4.4 × 3.0 cm

137

WB 31 A21

表胸针

1921 年
卡地亚巴黎
铂金
一颗玫瑰式切割钻石，单面切割和旧式切割钻
石
切面和圆柱形切割蓝宝石
黑玛瑙
黑色、白色珐琅
黑色波纹丝绸吊坠带（里金斯绸）
7.9 × 2.0 × 0.4 厘米

售予珍妮·浪万夫人。

此表为积家圆形机芯，日内瓦波
纹装饰，镀铑，8个调校项目，19颗宝
石轴承，瑞士杠杆式擒纵结构，双金属
平衡摆轮，宝玑摆轮游丝。

法国服装设计师珍妮·浪万
（1867～1946年）自1922年成为卡地
亚的客人，她曾担任1925年在巴黎举
行的艺术装饰博览会"时装与配饰"部
分负责人。在此次博览会上，路易·卡
地亚毗邻好友珍妮·浪万的"高级服饰
亭"设立了自己的主展位。

137

WB 31 A21

Watch-brooch

Cartier Paris, 1921
Platinum
One rose-cut diamond, single- and old-cut
diamonds
Faceted and calibré-cut sapphires
Onyx
Black and white enamel
Black moiré pendant ribbon (*régence*)
7.9 × 2.0 × 0.4 cm

Round LeCoultre movement, Côtes
de Genève decoration, rhodium-
plated, 8 adjustments, 19 jewels, Swiss
lever escapement, bimetallic balance,
Breguet balance spring.

Sold to Madame Jeanne Lanvin

French couturiere, Jeanne Lanvin,
(1867-1946), who had been a Cartier
client since 1922, was the head of the
"Fashion and Fashion accessories" section at the Exposition Internationale
des Arts Décoratifs, which was held in
Paris in 1925. During this exhibition,
Louis Cartier set up his main stand
near his friend Jeanne Lanvin in the
"Pavillon de l'Elegance".

138

BT 44 A21

手链

1921 年
卡地亚巴黎
铂金
圆形旧式切割、单面切割和玫瑰式切割钻石, 凸
圆形祖母绿和圆柱形切割祖母绿
天然珍珠
珊瑚（最初是两个黑玛瑙环, 而非珊瑚）
长度　17.0 厘米

售予黛丝·法罗斯夫人。

黛丝·法罗斯夫人是杜克·德斯卡
兹和伊莎贝拉·辛格（胜家缝纫机女继
承人）之女, 被20世纪20年代和30年
代的杂志誉为"全球最优雅的女人"。
的确, 这位巴黎社交名媛拥有两大出
类拔萃的品质：品味和大胆。如果没有
这两大品质, 她的优雅也会不为人所
瞩目。也正是这两大品质, 为她赢得了
时尚先锋的声望, 尤其是在1933年到
1935年间她担任著名时尚杂志《哈泼
时尚芭莎》的巴黎撰稿人期间（参见VC
85 A13, NE 28 A36, EG 28 A63）。

138

BT 44 A21

Bracelet

Cartier Paris, 1921
Platinum
Round old-, single- and rose-cut diamonds
Emerald cabochons and calibré-cut
emeralds
Natural pearls
Coral
Originally made with two rings of onyx
rather than coral.
Length 17.0 cm

Sold to Mrs. Daisy Fellowes

Daughter of the Duc Descaze and
Isabelle Singer (heiress to the Singer
sewing machine fortune), Daisy Fellowes was often named "the world's
most elegant woman" by magazines
of the 1920s and '30s. It is also true to
say that this Parisian socialite boasted
two qualities without which elegance
tends to be overlooked: taste and
audacity. These qualities earned her a
firm reputation as a leader of fashion,
especially when she became Paris
correspondent, from 1933 to 1935, for
the influential Harper's Bazaar.

(see VC 85 A13, NE 28 A36, EG 28
A63)

139
RG 23 A22

戒指

1922 年
卡地亚巴黎
铂金
圆形旧式切割钻石
一颗凸圆形祖母绿（重 2.31 克拉）
凸圆形珊瑚
黑玛瑙
2.40 × 2.10 × 0.60 厘米

139
RG 23 A22

Ring

Cartier Paris, 1922
Platinum
Round old-cut diamonds
One 2.31 carat emerald cabochon
Coral cabochons
Onyx
2.40 × 2.10 × 0.60 cm

140

CL 138 A22

胸针

1922 年
卡地亚巴黎
铂金
圆形旧式切割，单面切割和玫瑰式切割钻石
梨形和方形祖母绿，凸圆形祖母绿和祖母绿珠(一
颗为刻槽式，其余均为雕花式)
圆柱形切割黑玛瑙
15.1 × 4.2 厘米

这枚胸针创作于1922年，采用了
1920年的一个棘爪式胸针上的两颗黑
玛瑙和钻石图案。黑玛瑙阴影底纹为
这款作品赋予了立体感。

140

CL 138 A22

Brooch

Cartier Paris, 1922
Platinum
Round old-, single- and rose-cut diamonds
Pear-shaped and square emeralds, emerald
cabochons and emerald beads (one fluted,
the other carved)
Calibré-cut onyxes
15.1 × 4.2 cm

Made in 1922, this brooch incorporates two onyx and diamond motifs
taken from a cliquet brooch of 1920.
The onyx shading lends volume to
the piece.

141

CL 259 A22

胸针吊坠

1922 年
卡地亚巴黎, 特别订制
铂金
一颗梨形钻石, 圆形旧式切割和玫瑰式切割钻
石
珊瑚, 黑玛瑙
13.09 × 6.09 厘米

花瓶可单独搭配软绳佩戴。1914
年, 卡地亚基于1913年的一款设计,
共制作了三款类似胸针。此款胸针于
1922年订制, 使用了客人提供的一个
珊瑚花瓶。

141

CL 259 A22

Brooch-pendant

Cartier Paris, special order, 1922
Platinum
One pear-shaped diamond, round old- and
rose-cut diamonds
Coral, onyx
13.09 × 6.09 cm

The vase can be worn alone on a little cord. Three similar brooches were
made for stock in 1914 based on a
1913 design. This one was made to
order in 1922, using a coral vase supplied by the client.

束发带

1922 年
卡地亚巴黎
铂金，金
圆形旧式切割、单面切割和玫瑰式切割钻石
珊瑚珠和珊瑚条
黑玛瑙圆串珠和黑玛瑙条
玳瑁
黑色珐琅
高度　3 厘米

142

HO 24 A22

Bandeau

Cartier Paris, 1922
Platinum, gold
Round old-, single- and rose-cut diamonds
Coral beads and batons
Onyx roundels and batons
Tortoiseshell
Black enamel
Height 3 cm

I43

NE 24 A23

吊坠

1923 年
卡地亚巴黎, 特别订制
铂金
单面切割钻石
两颗凸圆形蓝宝石（其中一颗由客人提供, 重
121.02 克拉）
一颗雕花祖母绿（重 32.20 克拉）和一颗刻槽式
祖母绿珠（重 16.82 克拉）
9.90 × 3.0 × 1.85 厘米

这颗凸圆形蓝宝石来自斯里兰
卡, 未经过热加工。

I43

NE 24 A23

Pendant

Cartier Paris, special order, 1923
Platinum
Single-cut diamonds
Two sapphire cabochons (one supplied by
the client, weighing 121.02 carats)
One carved 32.20 carat emerald and one
fluted 16.82 carat emerald bead
9.90 × 3.0 × 1.85 cm

The sapphire cabochons are certified
as being from Sri Lanka and never
having undergone heat treatment.

带式胸针

1923 年
卡地亚巴黎
铂金
圆形旧式切割和单面切割钻石
凸圆形祖母绿
无色水晶
圆柱形和花式切割黑玛瑙
8.60 × 5.00 × 1.04 厘米

144
CL 66 A23

Belt brooch

Cartier Paris, 1923
Platinum
Round old- and single-cut diamonds
Emerald cabochons
Rock crystal
Calibré- and fancy-cut onyxes
8.60 × 5.00 × 1.04 cm

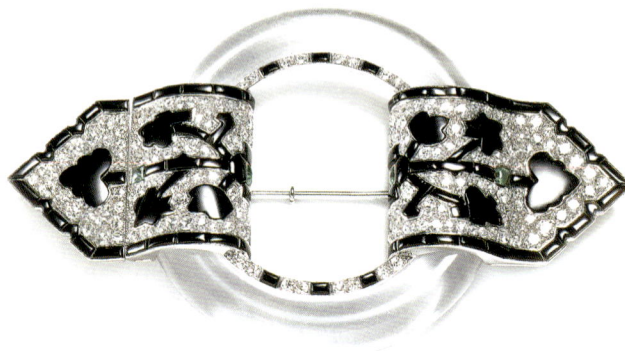

冠冕

1924 年
卡地亚纽约
铂金
一颗天然珍珠（重约 51 格令）
圆形旧式切割钻石，两颗梨形钻石
中心高度 5.25 厘米

来源于多丽丝·杜克遗产。

多丽丝·杜克（1912～1993年）是
美国烟草公司和杜克电力公司的创始
人詹姆斯·布加南·杜克和其第二任妻
子娜娜琳·霍尔特·因曼的女儿及唯一
继承人。她热衷于慈善事业，建立了
独立爱滋病基金会，也就是现在的多
丽丝·杜克基金会。她从她母亲娜娜
琳·杜克那里继承了这个冠冕。

145

HO 28 A24

Bandeau

Cartier New York, 1924
Platinum
A natural pearl weighing approximately 51 grains
Round old-cut diamonds, two pear-shaped diamonds
Height at centre 5.25 cm

Provenance: the Estate of Doris Duke

Daughter and only heiress of James Buchanan Duke – the founder of the American Tobacco Company and the Duke Power Company -, and his second wife, Nanaline Holt Inman, Doris Duke (1912-1993) blended her life passions with philanthropic causes. She established a foundation called Independent Aids, later to become the Doris Duke Foundation. She inherited this tiara from her mother, Nanaline Duke.

146

BT 90 A24

手链

1924 年
卡地亚巴黎
金，铂金
圆形旧式切割、单面切割和玫瑰式切割钻石
雕花珊瑚
螺钿
黑色珐琅
长度 18.0 厘米

此手链是卡地亚在1925年著名的"现代工业装饰艺术国际展览会"上的展品之一。

146

BT 90 A24

Bracelet

Cartier Paris, 1924
Gold, platinum
Round old-, single- and rose-cut diamonds
Carved coral
Mother-of-pearl
Black enamel
Length 18.0 cm

This bracelet was one of the items displayed by Cartier at the famous 1925 *Exposition Internationale des Arts Décoratifs et Industriels Modernes*.

147

BT 05 A25

手链

1925 年
卡地亚巴黎
铂金
方形、圆形旧式切割和单面切割钻石
珊瑚
黑玛瑙
长度　18.0 厘米

此手链是卡地亚在1925年著名的
"现代工业装饰艺术国际展览会"的展
品之一。

147

BT 05 A25

Bracelet

Cartier Paris, 1925
Platinum
Square, round old- and single-cut diamonds
Coral
Onyx
Length 18.0 cm

This bracelet was one of the items
displayed by Cartier at the famous
1925 *Exposition Internationale des Arts
Décoratifs et Industriels Modernes.*

148

CL 06 A25

水果碗胸针

1925 年
卡地亚巴黎
铂金, 金
梨形、菱形和方形单面切割和长阶梯形切割钻石
一颗凸圆形祖母绿和两颗凸圆形红宝石
刻槽式黑玛瑙
黑色珐琅
5.21 × 3.40 × 1.0 厘米

售予威廉·K·范德比尔特夫人 (参见CL 92 A05, CL 244 A25, CL 258 A22)。

中心位置原本镶嵌的是一颗雕花祖母绿。以黑玛瑙表现阴影, 手法高超精到, 强化了作品的透视感。长阶梯形切割在20世纪20年代盛极一时。卡地亚自20世纪早期就已开始使用此技法。

148

CL 06 A25

Fruit bowl brooch

Cartier Paris, 1925
Platinum, gold
Pear-, lozenge-shaped and square single- and baguette-cut diamonds
One emerald cabochon and two ruby cabochons
Fluted onyx
Black enamel
5.21 × 3.40 × 1.0 cm

A carved emerald was initially set in the middle. The sophisticated use of onyx to suggest shading enhances the effect of perspective. The baguette-cut technique, highly fashionable in the 1920s, had been used by Cartier since the early 1910s.

Sold to Mrs. William K. Vanderbilt (see CL 92 A05, CL 244 A25, CL 258 A22)

149

CL 262 A25

胸针

1925 年
卡地亚纽约, 特别订制
铂金
一颗枕形钻石 (重 3.83 克拉), 一颗梨形钻石,
圆形旧式切割钻石
一颗凸圆形祖母绿 (重 15.12 克拉)
珊瑚
黑色珐琅
4.80 × 3.80 × 1.60 厘米

149

CL 262 A25

Brooch

Cartier New York, special order, 1925
Platinum
One cushion-shaped 3.83 carat diamond, one pear-shaped diamond, round old-cut diamonds
One 15.12 carat emerald cabochon
Coral
Black enamel
4.80 × 3.80 × 1.60 cm

150

CL 50 C25

胸针

约 1925 年
卡地亚纽约
铂金，金
一颗梨形钻石，圆形旧式切割、单面切割和玫瑰
式切割钻石
黑色珐琅
7.77 × 1.98 厘米

150

CL 50 C25

Brooch

Cartier New York, c.1925
Platinum, gold
One pear-shaped diamond, round old-,
single- and rose-cut diamonds
Black enamel
7.77 × 1.98 cm

151

AN 14 C25

翠鸟

约 1925 年
卡地亚
无色水晶（基座），玛瑙（翠鸟），玉
银镀金
凸圆形蓝宝石（眼睛）
7.5 × 12.0 × 8.0 厘米

151

AN 14 C25

Kingfisher sculpture

Cartier, c. 1925
Rock crystal (base), agate (kingfisher), jade
Silver-gilt
Sapphire cabochons (eyes)
7.5 × 12.0 × 8.0 cm

152

BT 54 A26

手链

1926 年
卡地亚纽约
铂金，金
圆形旧式切割和单面切割钻石
镶钻石座红宝石饰珠
无色水晶（圆环和花瓶）
黑玛瑙（链节）
黑色珐琅
长度　18.6 厘米

152

BT 54 A26

Bracelet

Cartier New York, 1926
Platinum, gold
Round old- and single-cut diamonds
Ruby beads studded with collet-set
diamonds
Rock crystal (rings and vase)
Onyx (links)
Black enamel
Length 18.6 cm

153

BT 118 A26

带式手链

1926 年
卡地亚巴黎，特别订制
铂金
长阶梯形切割、法式切割、圆形旧式切割及单面
切割钻石
阶梯切割祖母绿
长度　19.0 厘米

153

BT 118 A26

Strap bracelet

Cartier Paris, special order, 1926
Platinum
Baguette-, French-, round old- and single-
cut diamonds
Step-cut emeralds
Length 19.0 cm

154

CL 293 A27

宝塔式胸针

1927 年
卡地亚巴黎
铂金
花式切割长阶梯形钻石，两颗三角形切割钻石
和一颗梨形钻石
3.75 × 1.52 厘米

154

CL 293 A27

Pagoda brooch

Cartier Paris, 1927
Platinum
Fancy-shaped baguette diamonds, two
triangular-cut diamonds and one pear-
shaped diamond
3.75 × 1.52 cm

155

CL 44 A27

庙宇胸针

1927 年
卡地亚巴黎为卡地亚纽约制作
铂金
一颗半月形钻石，长阶梯形切割和方形钻石
4.0 × 1.5 厘米

155

CL 44 A27

Temple brooch

Cartier Paris for New York, 1927
Platinum
One half-moon diamond, baguette-cut and
square diamonds
4.0 × 1.5 cm

156

TC 05 A27

旅行盥洗套件

1927 年
卡地亚伦敦
银镀金
半透明棕粉色、白色珐琅
象牙，玳瑁
玻璃
皮革
9.4 × 22.2 × 15.3 厘米

中央盒盖嵌一面镜子，内部装配
一个发梳和衣服刷，两个玻璃香精瓶，
一个带盖圆罐，两个唇膏架，一个指甲
锉和两个其他工具。另有一个隔层，内
设皮盒，装有三把钥匙和一把玳瑁梳。

156

TC 05 A27

Toilet travel set

Cartier London, 1927
Silver-gilt
Translucent brownish pink enamel over
guilloché ground, white enamel
Ivory, tortoiseshell
Glass
Leather
9.4 × 22.2 × 15.3 cm

The central cover with a mirror, the
interior fitted with a hair and a cloth
brush, two glass perfume flasks, a
circular pot with cover, two lipstick
holders, a nail file and two further
tools, a further compartment containing a leather box for three keys and a
tortoiseshell comb.

157

NE 14 A29

长项链

1928 年卡地亚巴黎（吊坠）
1929 年卡地亚巴黎（项链）
铂金
圆形旧式及单面切割钻石
项链总长度　79.0 厘米

　　此款长项链可被分成两条手链和
一条颈链，这是20世纪30年代流行的
佩戴方法。而今，这种长项链已经不再
流行。

157

NE 14 A29

Sautoir

Cartier Paris, 1928 (pendant) and 1929
(chain)
Platinum
Round old- and single-cut diamonds
Total length of the necklace 79.0 cm

This sautoir-type necklace could be
broken down into two bracelets and
a collar-type necklace. This is the way
it was worn in the 1930s, once the
fashion for long, sautoir-type chains
had passed.

158

BT 93 A28

带式手链

1928 年
卡地亚巴黎, 特别订制
铂金
面弧形及圆柱形切割蓝宝石, 长阶梯形切割、圆
形旧式切割及单面切割钻石
长度　17.00 厘米

158

BT 93 A28

Strap bracelet

Cartier Paris, special order, 1928
Platinum
Buff-top and calibré-cut sapphires,
baguette-, round old- and single-cut
diamonds
Length 17.00 cm

化妆盒

1928 年
卡地亚巴黎
金，铂金
黑色珐琅（盒身）
玫瑰式切割钻石，黑色珐琅斑点（豹身）
雕刻祖母绿，半透明绿色珐琅和圆柱形切割红
宝石（柏树）
两颗方形切割祖母绿和长阶梯形切割钻石（簧销）
10.85 × 5.5 × 1.75 厘米

盒内设一面化妆镜、两个隔层（一
个粉饼匣，一个烟匣）、一个唇膏架。

159

VC 08 A28

Vanity case

Cartier Paris, 1928
Gold, platinum
Black enamel (case)
Rose-cut diamonds, black enamel spots
(panther)
Carved emeralds, translucent green enamel
and calibré-cut rubies (cypresses)
Two carré-cut emeralds and a baguette-cut
diamond (thumb-piece)
10.85 × 5.5 × 1.75 cm

The interior fitted with a mirror, two
compartments, one for powder, the
other for cigarettes, a lipstick holder.

160

JA 05 A28

腰带扣

1928 年
卡地亚巴黎
金
雕花翡翠
绿松石
凸圆形蓝宝石
蓝色珐琅
深蓝色波纹丝绸腰带
3.37 × 6.78 × 0.5 厘米

160

JA 05 A28

Belt buckle

Cartier Paris, 1928
Gold
Carved jade
Turquoises
Sapphire cabochons
Blue enamel
Dark blue moiré belt
3.37 × 6.78 × 0.5 cm

161

CH 05 A28

烟管

1928 年
卡地亚巴黎
铂金
玫瑰式切割钻石
凸圆形祖母绿
墨玉
珊瑚
9.0 × 1.3 厘米

161

CH 05 A28

Cigarette holder

Cartier Paris, 1928
Platinum
Rose-cut diamonds
Emerald cabochons
Jet
Coral
9.0 × 1.3 cm

162

BT 79 A29

带式手链

1929 年
卡地亚纽约，特别订制
铂金
切面圆柱形切割红宝石，长阶梯形切割、明亮式
切割和圆形旧式切割钻石
长度　17.05 厘米

162

BT 79 A29

Strap bracelet

Cartier New York, special order, 1929
Platinum
Faceted and calibré-cut rubies, baguette-,
brilliant- and round old-cut diamonds
Length 17.05 cm

BT 97 A29

带式手链

1929 年
卡地亚纽约，特别订制
铂金
阶梯切割、改良长阶梯形切割钻石，明亮式切割
及单面切割钻石，子弹形钻石
长度　18.06 厘米

此款手链是各种类型切割钻石精
细、复杂组合的完美呈现，也是当时流
行的代表款式。

163

BT 97 A29

Strap bracelet

Cartier New York, 1929
Platinum
Step-, modified- and baguette-cut
diamonds, brilliant- and single-cut
diamonds, bullet-shaped diamonds
Length 18.06 cm

A fine example of the subtle, complex
play of various cuts of diamond, typical of a style then in vogue.

164

CL 76 A29

虎形胸针

1929 年
卡地亚巴黎
铂金
单面切割钻石
花式切割黑玛瑙（条纹）
1.6 × 3.3 厘米

164

CL 76 A29

Tiger brooch

Cartier Paris, 1929
Platinum
Single-cut diamonds
Fancy-cut onyxes (stripes)
1.6 × 3.3 cm

165

PB 01 A29

粉盒

1929 年
卡地亚巴黎
金
绿松石
青金石
5.4 × 5.4 × 1.5 厘米

铰链式盒盖，内嵌化妆镜。

165

PB 01 A29

Powder compact

Cartier Paris, 1929
Gold
Turquoise
Lapis lazuli
5.4 × 5.4 × 1.5 cm

The interior of the hinged cover fitted with a mirror.

雪茄盒

1929 年
卡地亚巴黎
金，铂金
长阶梯形切割钻石
珊瑚，螺钿
黑色珐琅
8.20 × 6.00 × 1.70 厘米

售予皮埃尔·卡地亚，他于20世纪50年代将其转赠于贞·杜桑（参见CL 289 A49）。

贞·杜桑（1887～1978年），杰出的时装设计师，有着罕见的优雅风度，主管卡地亚高级珠宝部。路易·卡地亚于1933年任命杜桑担任此职，以便自己能够投身于其他事务。两人的友谊始于1918年。自此以后，贞·杜桑对路易·卡地亚的影响力逐步增长。在从1933年到20世纪60年代的整个任职期间，这位拥有杰出魅力和创造力的女性事实上负责了整个卡地亚巴黎的高级珠宝设计。

盒盖装饰首字母缩写"JT"（贞·杜桑）。

166

PB 23 A29

Cigarette case

Cartier Paris, 1929
Gold, platinum
Baguette-cut diamonds
Coral, mother-of-pearl
Black enamel
8.20 × 6.00 × 1.70 cm

Initials JT (Jeanne Toussaint) on the cover.

Sold to Pierre Cartier who presented it to Jeanne Toussaint in the 50's.

Jeanne Toussaint (1887-1978) a great friend of fashion designers and a woman of rare elegance herself, henceforth headed Cartier's Fine Jewellery department. It was Louis Cartier who appointed Toussaint to do the job in 1933, when he wanted to relieve himself of certain duties. She had been his friend since 1918, and her influence steadily grew. From her appointment in 1933 to the 1960s, this woman of outstanding charm and creativity would in effect be responsible for high jewellery design at Cartier Paris.

(see CL 289 A49)

167

BT 13 A30

手链

1930 年
卡地亚巴黎
铂金，金
圆形旧式切割和单面切割钻石
刻槽式珊瑚珠
黑玛瑙
黑色珐琅
长度　20.5 厘米

每一颗珊瑚珠上镶嵌一颗钻石。
这款作品创作于1914年，其黑玛瑙豹
纹和铺镶钻石至今仍是卡地亚的标志
图案之一。

167

BT 13 A30

Bracelet

Cartier Paris, 1930
Platinum, gold
Round old- and single-cut diamonds
Fluted coral beads
Onyx spots
Black enamel
Length 20.5 cm

Each bead of coral in the central stand
is studded with a collet-set diamond.
Created in 1914, the panther-skin
onyx and diamond pavé setting has
remained one of Cartier's signature
motifs.

手镯

1930 年
卡地亚巴黎
铂金
圆形旧式切割及单面切割钻石
47 颗半圆形无色水晶
直径　7.05 厘米

售予格洛利亚·斯旺森（参见BT 27 A30）。

当时格洛利亚·斯旺森刚拍完她早期的有声电影（如《寡妇》和《钓金龟》），事业处于高峰阶段，并且刚与迈克尔·法玛尔完婚（她六个丈夫中的第三任）。她购自卡地亚的这对灵活柔韧、璀璨闪亮的手镯不仅书写了珠宝史的篇章，也为电影史增添了光芒。她在之后出演的至少两部电影中，都佩戴了这款手镯，如《完美的谅解》（1932年）以及《日落大道》（1950年）。

168

BT 28 A30

Bracelet

Cartier Paris, 1930
Platinum
Round old- and single-cut diamonds
Forty-seven rock crystal half-disks
Diameter 7.05 cm

Sold to Gloria Swanson. The star was then at the height of her fame, following her success in the early "talkies" (with movies such as *What a Widow!* and *Indiscreet*), and had just married Michael Farmer, the third of her six husbands. The pair of flexible, dazzling bracelets that she bought from Cartier are not only part of the history of jewellery but also part of the history of the movies, since she wore them in at least two films: *Perfect Understanding (1932)* and *Sunset Boulevard (1950)*.

(see BT 27 A30)

169

BT 27 A30

手镯

1930 年
卡地亚巴黎
铂金
长阶梯形切割、圆形旧式切割和单面切割钻石
30 颗半圆形无色水晶和 60 颗无色水晶珠
直径　7.85 厘米

售予格洛利亚·斯旺森（参见BT 28
A30）。

169

BT 27 A30

Bracelet

Cartier Paris, 1930
Platinum
Baguette-, round old- and single-cut
diamonds
Thirty rock crystal half-disks and sixty rock
crystal beads
Diameter 7.85 cm

Sold to Gloria Swanson

(see BT 28 A30)

170

CL 209 A30

喷泉胸针

1930 年
卡地亚巴黎，特别订制
铂金
一颗半月形钻石，长阶梯形切割、方形切割和圆
形旧式切割钻石
3.5 × 2.0 厘米

170

CL 209 A30

Fountain brooch

Cartier Paris, special order, 1930
One half-moon diamond, baguette-,
square- and round old-cut diamonds
3.5 × 2.0 cm

171

CDB 15 A30

可视机芯时钟
（仿积家 *101* 机芯）

1930 年
卡地亚纽约
黄铜
16.0 × 4.50 厘米

饰板上镌刻：致亨利·福特，感谢您创造了现代工业效率及建立迪尔伯恩博物馆——卡地亚，1930年1月。

积家长方形带截角机芯，8个调校项目，18颗宝石轴承，瑞士杠杆式擒纵结构，双金属平衡摆轮。

此时钟是1929年积家公司设计的101机芯原型的放大版。它是基于一个展示两层式机芯的模型而制作的。正如镌刻文字所示，此时钟为卡地亚纽约赠与亨利·福特之礼物。1992年，当它出现在亨利·福特博物馆和格林菲尔德镇（位于密歇根州的迪尔伯恩）的苏富比拍卖会时，卡地亚将其购回。

171

CDB 15 A30

Clock with apparent movement imitating the LeCoultre caliber 101

Cartier New York, 1930
Gilded brass
16.0 × 4.50 cm

Engraved on a plaque: To Henry Ford, a tribute of admiration to him as creator of modern industrial efficiency and of the Dearborn Museum from Cartier, January 1930.

Rectangular LeCoultre Duoplan movement with cut corners, 8 adjustments, 18 jewels, Swiss lever escapement, bimetallic balance.

This clock is a faithful enlargement of the caliber 101 movement devised by LeCoultre & Co. in 1929. It was based on a presentation model used to explain the two-level movement. As the inscription indicates, this clock was given by Cartier New York to Henry Ford. The Collection purchased it at Sotheby's in 1992 when it was placed on sale by The Henry Ford Museum and Greenfield Village (Dearborn, Michigan).

172

PB 09 A30

粉盒

1930 年
卡地亚巴黎为卡地亚伦敦制作
银，银镀金
玫瑰式切割钻石
珊瑚
黑色漆面
5.1 × 5.1 × 1.8 厘米

172

PB 09 A30

Powder box

Cartier Paris for Cartier London, 1930
Silver, silver gilt
Rose cut diamonds
Coral
Black lacquer
5.1 × 5.1 × 1.8 cm

173

PB 32 C30

粉盒

约 1930 年
卡地亚巴黎为卡地亚伦敦制作
铂金，金，银
玫瑰式切割钻石
珊瑚
黑色漆面
8.5 × 9.77 厘米

内嵌化妆镜。

173

PB 32 C30

Powder box

Cartier Paris, made for Cartier London, c. 1930
Platinum, gold, silver
Rose-cut diamonds
Coral
Black lacquer
8.5 × 9.77 cm

The interior fitted with a mirror.

烟盒

1930 年
卡地亚巴黎
金
长阶梯形切割镶钻按钮
青金石，内嵌绿松石
1.9 × 8.7 × 5.6 厘米

这一色彩搭配源自古埃及（参见朱迪·鲁德目录《卡地亚 1900～1939 年》，第210页）。

174

CC 47 A30

Cigarette case

Cartier Paris, 1930
Gold
Baguette-cut diamond-set push-piece
Lapis lazuli and turquoise inlay
1.9 × 8.7 × 5.6 cm

This colour combination derives from Ancient Egypt .
(see catalogue Judy Rudoe - Cartier 1900-1939, p. 210)

175

EB 31 C30

晚装包

约 1930 年

卡地亚

铂金

玫瑰式切割钻石

凸圆形祖母绿

珊瑚

天鹅绒

21.0 × 15.5 × 3.7 厘米

来源于杰奎琳·肯尼迪·欧纳西斯（参见苏富比纽约拍卖行目录，656号，2005年2月17日）。

175

EB 31 C30

Evening bag

Cartier, c. 1930

Platinum

Rose-cut diamonds

Emerald cabochons

Coral

Velvet

21.0 × 15.5 × 3.7 cm

Provenance: Jackie Kennedy-Onassis (see catalogue Sotheby's New York, lot 656, February 17, 2005)

RG 01 A33

戒指

1933 年
卡地亚纽约，特别订制
铂金
半月形和单面切割钻石
珊瑚
黑玛瑙
2.30 × 2.65 × 2.08 厘米

售予爱德华·F·赫顿夫人。

美国女实业家马乔里·梅瑞威瑟·波斯特（1887～1973年）是宝氏谷物公司继承人，同时也是一位热心的收藏家和慈善家。她在华盛顿建立了山林博物馆与花园，展出其所收藏的工艺品、绘画、家具和织物。1905年，她嫁给爱德华·本内特·克劳斯，后于1918年离异。同年，她嫁给爱德华·F·赫顿。1935年，她再披嫁衣，与美国驻俄大使约瑟夫·戴维联姻。

176

RG 01 A33

Ring

Cartier New York, special order, 1933
Platinum
Half-moon and single-cut diamonds
Coral
Onyx
2.30 × 2.65 × 2.08 cm

Sold to Mrs. Edward F. Hutton

The American businesswoman Marjorie Merriweather Post (1887-1973) was heir to the Post cereal firm. A keen collector and philanthropist, she founded Hillwood Museum and Gardens in Washington, D.C., where her collection of objets d'art, paintings, furniture, and textiles is displayed. She married Edward Bennet Close in 1905, but the couple divorced in 1918, and that same year she married Edward F. Hutton. In 1935 she wed Joseph Davies, U.S. ambassador to Russia.

CL 239 C33

胸针

约 1933 年
卡地亚伦敦, 特别订制
铂金
圆形旧式切割钻石
切面圆柱形切割蓝宝石
一颗长方柱形切面紫水晶
4.03 × 3.09 厘米

来源于雅克·卡地亚家族。

胸针上的宝石和家庭成员有着特殊的联系：紫水晶代表的是母亲；四个方形饰物标志着这对夫妻的四个孩子；圆柱形切割蓝宝石镶边则代表着父亲（这也是雅克·卡地亚的诞生石）。

177

CL 239 C33

Brooch

Cartier London, special order, c. 1933
Platinum
Round old-cut diamonds
Faceted and calibré-cut sapphires
One rectangular faceted amethyst
4.03 × 3.09 cm

Provenance: Jacques Cartier family

The stones in this brooch have specific symbolic associations to the family: the amethyst represents the mother, the four square motifs symbolize the couple's four children, and the calibré-cut sapphire borders represent the father (Jacques' birthstone).

178

EG 04 A34

夹式耳钉（一对）

1934 年
卡地亚伦敦
铂金
心形、三角形切割、方形切割和长阶梯形切割钻
石
1.50 × 2.00 厘米

178

EG 04 A34

Pair of ear clips

Cartier London, 1934
Platinum
Heart-shaped, triangular-, square- and
baguette-cut diamonds
1.50 × 2.00 cm

179

BT 70 A34

手镯（可拆为胸针）

1934 年
卡地亚巴黎
铂金
长阶梯形明亮式切割钻石
雕刻无色水晶
直径　7.50 厘米

中央饰物可以从手镯上取下用作
胸针。

179

BT 70 A34

Bangle with clip brooch

Cartier Paris, 1934
Platinum (osmior)
Baguette- and brilliant-cut diamonds
Carved and engraved rock crystal
Diameter 7.50 cm

The central motif can be detached
from the bracelet and worn as a
brooch on an osmior armature.

180

PE 11 A34

十字架吊坠

1934 年及 1949 年
卡地亚伦敦, 特别订制
十字架 (1934 年)
金, 银
玫瑰式切割、圆形旧式切割及单面切割钻石, 菱形和方形钻石
凸圆形蓝宝石, 一颗星彩蓝宝石, 切面圆柱形切割蓝宝石
圆形切面红宝石
珍珠钮
一颗蛋白石, 两颗仿蛋白石
圆形切面黄水晶
圣灵之鸽 (1949 年)
金
圆形旧式切割、单面切割和玫瑰式切割钻石, 一颗梨形钻石
一颗星彩蓝宝石 (重 17.20 克拉), 椭圆形切面蓝宝石, 一颗凸圆形蓝宝石, 一颗方形蓝宝石, 一颗切割祖母绿
切面及圆柱形切割红宝石
一颗珍珠
月长石
一颗蛋白石
25.30 × 7.20 × 1.45 厘米

180

PE 11 A34

Crucifix pendant

Cartier London, special orders, 1934 and
1949
Crucifix (1934)
Gold, silver
Rose-, single- and round old-cut diamonds,
lozenge-shaped and square diamonds
Sapphire cabochons, one star sapphire,
faceted and calibré-cut sapphires
Round faceted rubies
Button pearls
One opal and two imitation ones
Round faceted topazes
Dove of the Holy Spirit (1949)
Gold
Round old-, single- and rose-cut diamonds,
one pear-shaped diamond
One 17.20 carat star sapphire, oval faceted
sapphires, one sapphire cabochon, one
square sapphire
One emerald-cut emerald
Faceted and calibré-cut rubies
One pearl
Moonstones
One opal
25.30 × 7.20 × 1.45 cm

金字塔胸针

1935 年
卡地亚巴黎，特别订制
铂金
圆形旧式切割钻石（每颗重约 4.20 克拉），长阶
梯形切割及单面切割钻石
4.18 × 4.62 × 1.55 厘米

　　两枚一样的胸针，最初的设计理
念是既可分开佩戴，又可合为一体佩
戴。之后经客人要求，其中一枚胸针镶
嵌于一只带黑漆装饰的白金手链上。

181

CL 63 A35

Pyramid clip brooch

Cartier Paris, special order, 1935
Platinum
Round old-cut diamonds (one weighing
approximately 4.20 carats), baguette- and
single-cut diamonds
4.18 × 4.62 × 1.55 cm

Two identical clip brooches were
originally made to be worn separately
or as a single piece; the client later
asked that one of the brooches be set
on a bracelet of white gold and black
lacquer.

182

HO 06 A36

冠冕

1936 年
卡地亚伦敦，特别订制
铂金
旧式单面切割钻石
雕花绿松石
中心高度　4.8 厘米

售予罗伯特·亨利·布兰德。

罗伯特·亨利·布兰德
(1878～1963年)，银行家和资深政府
公务人员，曾为瑞德集团和劳埃德集
团总监。

182

HO 06 A36

Tiara

Cartier London, special order, 1936
Platinum
Old- and single-cut diamonds
Carved turquoises
Height at centre 4.8 cm

Sold to The Honorable Robert Henry Brand

Banker and senior civil servant, Robert Henry Brand (1878–1963) was a director of Lazard's and Lloyd's.

183

NE 32 A36

项链

1936 年
卡地亚伦敦
铂金
长阶梯形切割方形钻石
66 颗枕形及椭圆形切面红宝石(总重 97.72 克拉)
长度 40.0 厘米

183

NE 32 A36

Necklace

Cartier London, 1936
Platinum
Baguette-cut and square diamonds
Sixty-six cushion-shaped and oval faceted
rubies, weighing 97.72 carats in total
Length 40.0 cm

184

JS 06 A36

项链和手链

1936 年
卡地亚伦敦，特别订制
铂金
圆形旧式切割和长阶梯形切割钻石
一颗祖母绿式切割橄榄石（重 63.48 克拉），22
颗形态各异的祖母绿切割橄榄石（枕形、圆形和
椭圆形，总重 134.97 克拉）
项链直径　15 厘米，前高　6.07 厘米
手链　17.0 × 3.02 厘米

这组珠宝使用客户提供的橄榄石
订制而成。黄绿色的橄榄石与钻石关
系密切，因为它是在一种被称为角砾
云母橄榄岩的火成岩中发现，而钻石
亦产自火成岩。橄榄石是英国国王爱
德华七世最钟爱的宝石。

184

JS 06 A36

Necklace and bracelet

Cartier London, special order, 1936
Platinum
Round old- and baguette-cut diamonds
One emerald-cut 63.48 carat peridot,
twenty-two emerald-cut peridots of various
shapes (cushion-shaped, round and oval)
weighing 134.97 carats in total
Necklace diameter 15 cm, front height
6.07 cm
Bracelet 17.0 × 3.02 cm

This set was made to order, using
the client's peridots. The yellow-
green peridot, also called olivine or
chrysolite, is closely associated with
diamonds because it is found in kim-
berlite, the igneous rock that carries
diamonds to the surface of the earth.
Peridot was Edward VII's favorite
stone.

胸针

1936 年
卡地亚巴黎
金，铂金
明亮式切割钻石
一颗三角形深色黄水晶，梨形和长阶梯形切割
浅色黄水晶
3.50 × 3.50 厘米

最初，两枚同样的胸针是佩戴在
一条由抛光黄金制成的圆形手链上。
但最后这两枚胸针被分开出售。

185

CL 137 A36

Clip brooch

Cartier Paris, 1936
Gold, platinum
Brilliant-cut diamonds
One triangular dark citrine, lighter pear-
shaped and baguette-cut citrines
3.50 × 3.50 cm

Initially, two identical brooches were
made to be worn on a hoop bracelet of polished yellow gold, but the
brooches were ultimately sold separately.

186

HO 14 A37

冠冕

1937 年
卡地亚伦敦
金，铂金
圆形旧式切割和长阶梯形切割钻石
一颗大号八角形祖母绿式切割深色黄水晶（重
62.35 克拉）
圆柱形切割和长阶梯形切割黄水晶，一颗六角
形黄水晶
直径　17.0 厘米
胸针　5.8 × 4.7 厘米

中央的饰物可以从冠冕上取下，
用作胸针（头向下）。

186

HO 14 A37

Tiara

Cartier London, 1937
Gold, platinum
Round old- and baguette-cut diamonds
One large octagonal emerald-cut 62.35
carat dark citrine
Calibré- and baguette-cut citrines, one
hexagonal citrine
Diameter 17.0 cm; clip-brooch 5.8 × 4.7 cm

The central motif can be detached
from the tiara and worn as a brooch,
pointing downward.

187

BT 48 A37

手镯

1937 年
卡地亚巴黎
铂金
一颗枕形旧式切割钻石（重 5.66 克拉），两颗圆
形旧式切割钻石（分别重 3.78 克拉和 3.58 克拉），
方形切割、长阶梯形切割、明亮式切割、圆形旧
式切割和单面切割钻石
5.7 × 7.0 × 2.6 厘米

187

BT 48 A37

Bangle

Cartier Paris, 1937
Platinum
One cushion-shaped old-cut 5.66 carat
diamond, two round old-cut diamonds
(3.78 and 3.58 carats), square-, baguette-,
brilliant-, round old- and single-cut
diamonds
5.7 × 7.0 × 2.6 cm

188

BT 01 A37

手链

1937 年
卡地亚巴黎
铂金，白金，银
圆形旧式切割和单面切割钻石
青金石珠
氧化银内嵌凸圆形绿松石
长度　20.3 厘米

188

BT 01 A37

Bracelet

Cartier Paris, 1937
Platinum, white gold, silver
Round old- and single-cut diamonds
Lapis lazuli beads
Turquoise cabochons on oxidized silver
Length 20.3 cm

189

CL 250 A37

胸针

1937 年
卡地亚伦敦
铂金
明亮式切割和长阶梯形切割钻石
9 颗阶梯切割圆形蓝宝石
3.71 × 3.66 厘米

189

CL 250 A37

Clip brooch

Cartier London, 1937
Platinum
Brilliant- and baguette-cut diamonds
Nine step-cut sapphires of circular shape
3.71 × 3.66 cm

190

CL 132 A37

兰花胸针

1937 年
卡地亚巴黎，特别订制
白金
切面花式切割紫水晶和海蓝宝石
宝石间点缀镶嵌淡蓝色和淡紫色珐琅
12.55 × 11.38 厘米

190

CL 132 A37

Orchid brooch

Cartier Paris, special order, 1937
White gold
Faceted fancy-cut amethysts and
aquamarines
Pale blue and mauve enamel studs
between the stones
12.55 × 11.38 cm

191

CL 267 A37

胸针

1937 年
卡地亚纽约，特别订制
铂金
枕形、长阶梯形切割和圆形旧式切割钻石
两颗水滴形祖母绿和一颗凸圆形祖母绿
5.46 × 4.9 × 1.1 厘米

售予艾文·柏林夫人。

艾琳·迈凯(1902～1988年)为美国邮政电报公司总裁克拉伦斯·迈凯之女，她是康斯托克银矿矿主约翰·迈凯的孙女。艾琳·迈凯于1926年嫁给著名美国作曲家艾文·柏林。她曾是活跃在媒体界的记者和作家。在第二次世界大战期间，她曾加入援助盟军保卫美国委员会，强烈反对德国的纳粹政权。

191

CL 267 A37

Clip brooch

Cartier New York, special order, 1937
Platinum
Cushion-shaped, baguette- and round old-
cut diamonds
Two emerald drops and one emerald
cabochon
5.46 × 4.9 × 1.1 cm

Sold to Mrs. Irving Berlin

Daughter of Clarence Mackay (head of the American Post and Telegraph Company) and grand-daughter of John Mackay (owner of the Comstock Lode silver mines), Ellin Mackay (1902–1988) married the famous American composer Irving Berlin in 1926. She was an active journalist and novelist. During the Second World War she joined the Committee to Defend America by Aiding the Allies, vehemently opposing the Nazi regime in Germany.

192

CL 184 A38

玫瑰胸针

1938 年
卡地亚巴黎
黄金, 玫瑰金, 铂金
一颗圆形旧式切割钻石, 玫瑰式切割钻石
雕花珊瑚
黑色珐琅
4.00 × 2.50 厘米

192

CL 184 A38

Rose clip brooch

Cartier Paris, 1938
Yellow gold, pink gold, platinum
One round old-cut diamond, rose-cut
diamonds
Carved coral
Black enamel
4.00 × 2.50 cm

193

WWL 71 A38

手镯式腕表

1938 年
卡地亚巴黎
金
切面方形和长方形黄水晶
表壳宽度　1.05 厘米
表链宽度　2.5 厘米

此表为积家403长方形机芯, 镀
铑, 15个宝石轴承, 瑞士杠杆式擒纵结
构, 双金属平衡摆轮, 扁平摆轮游丝。

193

WWL 71 A38

Bracelet-watch

Cartier Paris, 1938
Gold
Faceted square- and rectangular-shape
citrines
Width of the case 1.05 cm; width of the
bracelet 2.5 cm

Rectangular LeCoultre caliber 403
Duoplan movement, rhodium-plated,
15 jewels, Swiss lever escapement,
bimetallic balance, flat balance spring.

194

BT 105 A39

手铐式手镯

1939 年
卡地亚巴黎
金
圆锥形紫水晶
圆柱形切割长方形黄水晶
直径　7.50 厘米

194

BT 105 A39

Handcuff bracelet

Cartier Paris, 1939
Gold
Conical amethyst cabochons
Rectangular calibré-cut citrines
Diameter 7.50 cm

195

BT 45 A39

手镯（可用作胸针）

1939 年
卡地亚巴黎
白金，铂金
明亮式切割钻石
梨形海蓝宝石
每枚夹子　6.6 × 2.8 厘米

195

BT 45 A39

Bracelets with clip brooches

Cartier Paris, 1939
White gold, platinum
Brilliant-cut diamonds
Pear-shaped aquamarines
Each clip 6.6 × 2.8 cm

196

RG 19 A41

梨形镶嵌宝石戒指

1941 年
卡地亚巴黎
铂金
明亮式切割钻石
四颗枕形蓝宝石（总重为 18.10 克拉）
3.23 × 2.45 × 2.06 厘米

196

RG 19 A41

Boule ring

Cartier Paris, 1941
Platinum
Brilliant-cut diamonds
Four cushion-shaped sapphires, weighing
18.10 carats in total
3.23 × 2.45 × 2.06 cm

197

CL 188 A41

翠鸟胸针

1941 年
卡地亚巴黎
铂金，金
明亮式切割和单面切割钻石
两颗叶形雕花祖母绿（总重 17.66 克拉）
切面圆柱形切割蓝宝石
两颗凸圆形红宝石
7.84 × 5.40 × 0.78 厘米

197

CL 188 A41

Kingfisher clip brooch

Cartier Paris, 1941
Platinum, gold
Brilliant- and single-cut diamonds
Two leaf-shaped carved emeralds, weighing
17.66 carats in total
Faceted and calibré-cut sapphires
Two ruby cabochons
7.84 × 5.40 × 0.78 cm

198

CL 236 A41

花朵胸针

1941 年
卡地亚巴黎，特别订制
铂金，白金
圆形旧式切割及单面切割钻石
长度　12.5 厘米

198

CL 236 A41

Flower clip brooch

Cartier Paris, special order, 1941
Platinum, white gold
Round old- and single-cut diamonds
Length 12.5 cm

199

CL 304 A42

花形胸针

1942 年
卡地亚巴黎
金，铂金
明亮式切割钻石
9 颗凸圆形祖母绿
7.4 × 5.5 × 1.7 厘米

199

CL 304 A42

Panache brooch

Cartier Platinum, 1942
Brilliant-cut diamonds
Nine emerald cabochons
7.4 × 5.5× 1.7 cm

200

CL 180 A43

月桂枝胸针

1943 年
卡地亚巴黎
金，铂金
明亮式切割钻石
4 颗凸圆形蓝宝石
9.0 × 4.50 厘米

200

CL 180 A43

Laurel clip brooch

Cartier Paris, 1943
Gold, platinum
Brilliant-cut diamonds
Four sapphire cabochons
9.0 × 4.50 cm

201

CL 191 A44

鸟形胸针

1944 年
卡地亚巴黎
金，铂金
明亮式切割和单面切割钻石
4 颗凸圆形祖母绿（总重 14.42 克拉）
椭圆形切面蓝宝石
两颗马眼凸圆形切割红宝石
9.00 × 7.05 厘米

201

CL 191 A44

Bird clip brooch

Cartier Paris, 1944
Gold, platinum
Brilliant- and single-cut diamonds
Four emerald cabochons, weighing 14.42
carats in total
Oval faceted sapphires
Two navette-shaped ruby cabochons
9.00 × 7.05 cm

202

CL 77 A44

自由鸟胸针

1944 年
卡地亚巴黎
金，铂金
玫瑰式切割钻石
一颗凸圆形蓝宝石
青金石
珊瑚
3.72 × 2.41 × 1.7 厘米

第二次世界大战期间，卡地亚设计了一系列以"笼中之鸟"为主题的艺术品，象征被占领的法国。当巴黎解放后，彩色的鸟儿也满心欢喜地冲出了牢笼。

202

CL 77 A44

Oiseau libéré brooch

Cartier Paris, 1944
Gold, platinum
Rose-cut diamonds
One sapphire cabochon
Lapis lazuli
Coral
3.72 × 2.41 × 1.7 cm

During the Second World War Cartier designed various pieces featuring caged birds, symbolizing occupied France. Once Paris was liberated, the patriotically coloured bird could burst from its cage in an explosion of joy.

203

CL 126 A47

鸟形胸针

1944 ～ 1947 年
卡地亚巴黎
錾刻金，铂金
圆形旧式切割、明亮式和单面切割钻石
一颗凸圆形切割红宝石（重 43.20 克拉）
一颗凸圆形切割祖母绿
5.6 × 3.02 × 1.6 厘米

此款胸针的原型为1944年的"天堂鸟"胸针，后于1947年去除其镶红宝石和凸圆形祖母绿錾刻黄金长尾翼。

203

CL 126 A47

Bird clip brooch

Cartier Paris, 1944-1947
Chased gold, platinum
Round old-, brilliant- and single-cut diamonds
One 43.20 carat ruby cabochon
One emerald cabochon
5.6 × 3.02 × 1.6 cm

This piece was originally a "bird of paradise" brooch made in 1944, until the removal in 1947 of its long, chased-gold tail with ruby and emerald cabochons.

204

RG 36 A45

戒指

1945 年
卡地亚巴黎
金
一颗方形阶梯切割祖母绿
8 颗椭圆形切面红宝石
2.80 × 2.25 × 1.72 厘米

204

RG 36 A45

Ring

Cartier Paris, 1945
Gold
One square step-cut emerald
Eight oval faceted rubies
2.80 × 2.25 × 1.72 cm

205

BT 123 C45

手镯

约 1945 年
卡地亚纽约
玫瑰金，铂金
圆形切面蓝宝石
圆形旧式切割钻石
高　6.75 厘米
直径　5.0 厘米

205

BT 123 C45

Bracelet

Cartier New York, c. 1945
Pink gold, platinum
Round faceted sapphires
Round old-cut diamonds
Height 6.75 cm; diameter 5.0 cm

BT 11 A45

管式手镯

1945 年
卡地亚巴黎
金，铂金
圆形旧式切割钻石
一颗凸圆形蓝宝石（重 23.37 克拉）
叶形雕花红宝石
长度　23.50 厘米

206

BT 11 A45

Gaspipe bracelet

Cartier Paris, 1945
Gold, platinum
Round old-cut diamonds
One 23.37 carat sapphire cabochon
Leaf-shaped carved rubies
Length 23.50 cm

蝴蝶胸针

1945 年
卡地亚巴黎
金，铂金
单面切割钻石
凸圆形祖母绿
雕花珊瑚
黑色珐琅
3.64 × 2.07 × 0.73 厘米

来源于若瑟蒂·狄。

法国女演员若瑟蒂·狄(1914～1978年)从5岁起就开始出现在电影荧屏上。她扮演的第一个广为人知的角色是在20世纪30年代下半期。她曾和众多导演合作，包括阿贝尔·冈斯(电影《勒格雷·波基雅》，1935年)，朱利恩·杜维威尔(电影《白天的男人》，1936年)以及马塞尔·帕尼奥尔(战时的合作伙伴，曾于1940年为她写了电影《掘井工的女孩》的剧本)。1946年，让·考克多邀请她与让·玛莱丝一起出演电影《美女与野兽》中的美女角色，并于1948年在电影《可怕的父母》中再次邀请其演出。1950年若瑟蒂·狄嫁给比利时商人莫里斯·苏威尔并从此退出影坛(参见EG 26 A52, CO 13 A52, BT 23 A72)。

207

CL 64 A45

Butterfly clip brooches

Cartier Paris, 1945
Gold, platinum
Single-cut diamonds
Emerald cabochons
Carved coral
Black enamel
3.64 × 2.07 × 0.73 cm

Provenance: Josette Day

The French actress Josette Day (1914–1978) started acting in the movies at the age of five. Her first major roles date from the latter half of the 1930s. She worked with directors such as Abel Gance (*Lucretia Borgia*, 1935), Julien Duvivier (*L'Homme du Jour*, 1936) and Marcel Pagnol (her partner during the war, who wrote the screenplay of *La Fille du Puisatier* for her in 1940). In 1946, Jean Cocteau asked her to play Beauty alongside Jean Marais in *La Belle et la Bête,* and then cast her in *Les Parents Terribles* in 1948. In 1950 she married Belgian businessman Maurice Solvay and retired from the movie industry.

(see EG 26 A52, CO 13 A52, BT 23 A72)

粉盒

1946 年
卡地亚纽约
凸圆形及圆柱形切割红宝石
金
7.66 × 1.42 × 6.58 厘米

售予费雯丽。

内设一个粉饼隔层和一面镜子。表面刻有字母组合GL、名字首字母V+L以及日期1952。

费雯丽（1913～1967年）是英国最杰出女演员之一。她曾就读于伦敦皇家戏剧艺术学院并出演19部影片。1940年，费雯丽因成功诠释电影《乱世佳人》中的斯嘉丽一角而获奥斯卡奖。1951年，她又因在电影《欲望号街车》中的出色表演而第二次获奥斯卡奖，向全世界证明了她的才华。

Powder case

Cartier New York, 1946
Gold
Cabochon and calibré-cut rubies
7.66 × 1.42 × 6.58 cm

The interior fitted with a powder compartment and a mirror. The cover engraved with the monogram GL, the initials V+L and the date 1952.

Sold to Vivien Leigh. Born Vivian Hartley (1913-1967), one of the most famous English actresses. She studied at the Royal Academy of Dramatic Art in London and played in 19 films. In 1940, Vivien Leigh won the Oscar for her phenomenal portrayal of heroine Scarlett O'Hara in *Gone with the Wind* and a second Oscar in 1951 for her performance in the movie *A streetcar named Desire*. She received a world-wide recognition for her talent.

209

VC 45 A47

化妆盒

1947 年
卡地亚伦敦
金
方形及圆柱形切割蓝宝石
凸圆形蓝宝石
14.2 × 6.55 × 4.35 厘米

售予一名意大利皇家成员。
内设一面镜子，四个翻盖隔层。

209

VC 45 A47

Vanity case

Cartier London, 1947
Gold
Square- and calibré- cut sapphires
Sapphire cabochons
14.2 × 6.55 × 4.35 cm

The interior fitted with a mirror and
four compartments.

Sold to a member of the Royal
family of Italy

210

NE 18 A49

棕榈树项链

1949 年
卡地亚巴黎，特别订制
编织和抛光金，铂金
单面切割钻石
镶有单面切割钻石底座的红宝石珠
宽度　17.0 厘米
棕榈树高度　11.4 厘米

此款项链所用宝石以及印度饰物
（镶嵌于两侧）均为客户所提供。

210

NE 18 A49

Palm-tree necklace

Cartier Paris, special order, 1949
Plaited and polished gold, platinum
Single-cut diamonds
Ruby beads studded with collet-set single-cut diamonds
Width 17.0 cm; height of palm tree 11.4 cm

This necklace was made from stones and two old Hindu motifs (on the sides) supplied by the client.

2II

CL 289 A49

箭袋胸针

1949 年
卡地亚巴黎
黄金，玫瑰金
白玛瑙
圆柱形切割红宝石
一颗长阶梯形切割钻石，玫瑰式切割钻石
4.85 × 2.75 厘米

售予贞·杜桑女士（参见 PB 23
A49）。

2II

CL 289 A49

Quiver pin-brooch

Cartier Paris, 1949
Yellow gold, pink gold
White onyx
Calibré-cut rubies
One baguette-cut diamond, rose-cut
diamonds
4.85 × 2.75 cm

Sold to Mlle Jeanne Toussaint

(see PB 23 A49)

2I2

LR 10 A49

汽油打火机

1949 年
卡地亚巴黎
银
玫瑰式切割钻石
珊瑚瓢虫
黑色漆面
3.85 × 3.49 × 1.49 厘米

2I2

LR 10 A49

Petrol lighter

Cartier Paris, 1949
Silver
Rose-cut diamonds
Coral ladybird
Black lacquer
3.85 × 3.49 × 1.49 cm

213

CDS 02 A49

支架式座钟

1949 年
卡地亚巴黎
铂金，金，银镀金
翡翠圆环
中央镶镂空珊瑚饰物，切面珊瑚珠
螺钿
玫瑰式切割钻石
直径　9.18 厘米

此钟为8日储存圆形机芯，镀铑，1个调校项目，11颗宝石轴承，瑞士杠杆式擒纵结构，单金属平衡摆轮。

213

CDS 02 A49

Desk clock with strut

Cartier Paris, 1949
Platinum, gold, gilded silver
Jade ring
Openwork carved coral plaque in the center, faceted coral beads
Mother-of-pearl
Rose-cut diamonds
Diameter 9.18 cm

Round 8-day movement, rhodium-plated, 1 adjustment, 11 jewels, Swiss lever escapement, monometallic balance.

214
EG 26 A52

克里奥耳夹式耳环（一对）

1952 年
卡地亚巴黎
铂金，白金
明亮式切割钻石
镶钻珊瑚
背面为黑色漆面
直径　3.68 厘米

售予若瑟蒂·狄（参见CL 64 A45,
CO 13 A52, BT 23 A72）。

214
EG 26 A52

Pair of Creole ear clips

Cartier Paris, 1952
Platinum, white gold
Brilliant-cut diamonds
Coral studded with collet-set diamonds
Black lacquer on the back
Diameter 3.68 cm

Sold to Josette Day

(see CL 64 A45, CO 13 A52, BT 23
A72)

215

CO 13 A52

狮首腕表

1952 年
卡地亚巴黎
金，铂金
雕花珊瑚
长阶梯形和明亮式切割钻石，单面切割及星形
切割钻石，玫瑰式切割钻石
两颗马眼形切割祖母绿（眼睛）
黑色珐琅
4.0 × 3.0 厘米

售予若瑟蒂·狄（参见CL 64 A45, EG 26 A52, BT 23 A72）。

此表为积家169圆形机芯，模拟日内瓦波纹装饰，镀铑，2个调校项目，18颗宝石轴承，瑞士杠杆式擒纵结构，双金属平衡摆轮，扁平摆轮游丝。

215

CO 13 A52

Lion's-head watch

Cartier Paris, 1952
Gold, platinum
Carved coral
Baguette-, brilliant-, single- and star-set-
diamonds, rose-cut diamond
Two navette-cut emeralds (eyes)
Black enamel
4.0 × 3.0 cm

Round LeCoultre caliber 169 movement, fausses Côtes de Genève decoration, rhodium-plated, 2 adjustments, 18 jewels, Swiss lever escapement, bimetallic balance, flat balance spring.

Sold to Josette Day

(see CL 64 A45, EG 26 A52, BT 23 A72).

手镯

1953 年
卡地亚巴黎
绞纹黄金, 黄金丝和抛光黄金, 白金
明亮式切割钻石
凸圆形绿松石
6.50 × 7.10 厘米

售予罗斯柴尔德家族成员 (参见
CL 249 A12, BT 108 A13)。

这只手镯以铰链打开。

216

BT 104 A53

Bangle

Cartier Paris, 1953
Twisted yellow gold, yellow gold wire and
polished yellow gold, white gold
Brilliant-cut diamonds
Turquoise cabochons
6.50 × 7.10 cm

This bracelet opens on a hinge.

Sold to a member of the Rothschild family

(see CL 249 A12, BT 108 A13)

217

CL 54 A53

蜻蜓胸针

1953 年
卡地亚巴黎
铂金，金
花式切割、祖母绿切割、明亮式切割、单面切割、
长阶梯形切割和玫瑰切割钻石
一颗凸圆形祖母绿
2 颗圆形切面红宝石和 10 颗圆柱形切割红宝石
10.03 × 6.06 厘米

配有镂空可颤动翅膀。每个翅膀
由一条弹簧连接于身体部分，可以微
微颤动。

217

CL 54 A53

Dragonfly clip brooch

Cartier Paris, 1953
Platinum, gold
Fancy-, emerald-, brilliant-, single-,
baguette- and rose-cut diamonds
One emerald cabochon
Two round faceted rubies and ten calibré-
cut rubies
10.03 × 6.06 cm

The tremored, openwork setting of
the wings, each attached to the body
by a spring, allows them to flutter at
the tiniest movement.

218

BT 26 A54

喀迈拉手镯

1954 年
卡地亚巴黎，特别订制
铂金，白金
马眼形和明亮式切割钻石
刻槽式珊瑚珠和面弧形凸圆形珊瑚
20.0 × 2.01 厘米

218

BT 26 A54

Chimera bangle

Cartier Paris, special order, 1954
Platinum, white gold
Marquise- and brilliant-cut diamonds
Fluted coral beads and buff-top coral
cabochons
20.0 × 2.01 cm

219

NE 31 A55

吊坠项链

1955 年
卡地亚巴黎
金丝和编织金
明亮式切割钻石
凸圆形绿松石
中央饰物长度 7.60 厘米

219

NE 31 A55

Bib necklace

Cartier Paris, 1955
Gold wire and plaited gold
Brilliant-cut diamonds
Turquoise cabochons
Length of the central motif 7.60 cm

让·考克多的院士剑

1955年
卡地亚巴黎
金，银
祖母绿，红宝石，钻石
象牙，黑玛瑙，蓝色珐琅
钢制剑身
长度 87.0 厘米

让·考克多(1889～1963年)，才华横溢的法国作家，创作了多部诗歌、小说以及画作。剑的护手盘描绘了奥菲斯的形象——奥菲斯的神话故事一直贯穿于让·考克多的诗歌中。剑柄的圆头设计为象牙七弦琴造型，镶嵌祖母绿(重2.84克拉)和两颗红宝石。剑柄圆柱形，环绕着仿古老戏院装饰风格的织物图案金饰，象征着悲剧。剑鞘上镌刻诗人的签名：名字的首字母和一颗星星——星星同样出现在较大的镶嵌钻石和红宝石的象牙饰物上。炭笔造型的剑套象征着他的画作，剑鞘上的饰物灵感来自让·考克多曾居住的皇宫。剑鞘尾部装饰着一只手握着的象牙珠——暗示着他的剧本《坏孩子》中以雪包裹着的宝石。剑身由一位西班牙友人提供，产于托莱多的一家兵器厂。祖母绿来自可可·香奈儿，钻石和红宝石来自法兰辛·威斯威尔。

路易十三统治时期的红衣主教黎塞留在1634年建立法兰西学院，其位于巴黎康迪码头法兰西研究院的一幢由建筑师勒沃(1612～1670年)所设计的建筑内。法兰西学院至今仍然是西方最顶级的研究机构之一，以规范和完善法国语言为己任。学院共有40名由同行推举的终生院士。院士需身着黑色院士袍，头戴绿色橄榄枝和双角帽，身披斗篷，手持一把由友人、仰慕者或家乡委员会所赠送的院士剑。在1931年到1974年期间，卡地亚共接到了23宗院士剑委任案。

法国著名诗人让·考克多是路易·卡地亚的朋友，两人的友谊始于20世纪20年代。这段恒久而忠诚的友谊一直延续到路易去世以后。1955年，让·考克多当选法兰西学院院士。他的朋友们一致决定，委任卡地亚为他打造一把院士剑。

和平街的珠宝商已经创制了23把类似的剑，每一把都凝聚着未来院士与珠宝艺术家之间亲密对话。让·考克多的院士剑，由诗人本人亲手绘制草图，毫无争议地成为了卡地亚工作坊最富创意的院士剑。这件优美的艺术品是诗人创作与永恒传奇的真正象征。

Jean Cocteau's Academician sword

Cartier Paris, 1955
Gold, silver
Emerald, rubies, diamond
Ivory, onyx, blue enamel
Steel blade
Length 87.0 cm

Jean Cocteau, French writer (1889-1963), he expressed his talent through poems, novels and numerous drawings.

The guard of the sword depicting the profile of Orpheus whose myth haunted the poet, the pommel designed as an ivory lyre set with an emerald weighing 2.84 carats and two rubies, the hilt designed as a column with a coiled gold fabric-like motif in the manner of antique theatre set decoration, symbolizing tragedy, on the sheath the signature of the poet: his initials and a star, the latter also appears above in a larger ivory motif set with diamonds and rubies, the guard of charcoal-pencil shape representing his graphic works, the sheath applied with a motif inspired by the gates of the Palais Royal where Jean Cocteau lived, the end of the sheath decorated with an ivory bead grasped by a hand, representing the stone coated by snow from his play *Les Enfants terribles*,

The blade was given by Spanish friends, which was originally from an arms factory in Toledo. The emerald was given by Coco Chanel, the diamond and rubies by Francine Weisweiller.

Swords for Academicians

The French Académie was founded under the reign of Louis XIII by the Cardinal de Richelieu in 1634. Located in the sumptuous building of the architect Le Vau (1612-1670) of the Institut de France Quai de Conti in Paris, it is still one of the most prestigious institutions in the Western world. Its function is to normalize and perfect the French language. The 40 members, appointed for life by their peers, wear a black ceremonial dress with green olive branches, a bicorne hat, a cape and a sword presented to the new Academician by friends, admirers or a committee from his or her home town. Between 1931 and 1974 Cartier received twenty-three commissions for academician swords.

Jean Cocteau was a friend of Louis Cartier since the 1920's, which resulted in a long and loyal relationship between the Poet and the firm even after the death of Louis. In 1955 when he was elected to the Académie Française, his friends joined together and naturally turned to commission his sword from Cartier.

The rue de la Paix jeweller had already crafted some twenty-three such swords, each one arising from an intimate dialogue between the future academician and the artist-jeweller. In this occasion the drawing for Cocteau's sword would be by the poet's own hand, resulting without any doubt in the most original and surprising of all produced in the Cartier workshops. This beautiful work of art is a veritable encapsulation of the poet's work and the legends that inhabit it.

221

CM 15 A56

单轴承魅幻时钟

1956 年
卡地亚巴黎
抛光织纹金，铂金
玫瑰式切割、明亮式切割和单面切割钻石
烟色石英钟盘
21 × 17 × 8.5 厘米

此钟为8日储存长方形机芯，镀金，瑞士杠杆式擒纵结构，双金属平衡摆轮，宝玑摆轮游丝。上弦轴柄和调校轴柄设于基座后。

221

CM 15 A56

Mystery clock with single axle

Cartier Paris, 1956
Polished and textured gold, platinum
Rose-, brilliant- and single-cut diamonds
Smoky quartz dial
21 × 17 × 8.5 cm

Rectangular 8-day movement, gold-plated, Swiss lever escapement, bime-tallic balance, Breguet balance spring. Arbor for winding movement and setting hands on back of base.

222

EB 25 A57

晚装手袋

1957 年
卡地亚纽约
金, 铂金
圆形切割、长阶梯形切割和明亮式切割钻石
雕花珊瑚
凸圆形祖母绿和蓝宝石
黑色天鹅绒
18.0 × 17.0 厘米

来源于玛莉亚·菲利克斯（参见1996年5月15日，日内瓦佳士得目录106号）。

玛莉亚·菲利克斯(1914～2002年)是墨西哥首屈一指的著名女演员，典型的拉丁美女。她的演艺生涯始于20世纪60年代末期。她曾出演多部著名墨西哥和法国电影，包括埃米里奥·费尔南德兹导演的电影《由恨而生的爱》(1946年)，让·雷诺瓦导演的《法国康康舞》(1955年)，以及路易斯·布努埃尔导演的《帕欧的火山》(1959年)。玛莉亚·菲利克斯十分钟情于蛇皮及鳄鱼皮等产品(参见NE 10 A68, NE 43 A75)。

222

EB 25 A57

Evening bag

Cartier New York, 1957
Gold, platinum
Round-, baguette- and brilliant-cut
diamonds
Carved coral
Emerald and sapphire cabochons
Black velvet
18.0 × 17.0 cm

Provenance: María Félix (see catalogue Christie's Geneva, lot 106, May 15, 1996)

Mexican diva and archetypal Latin femme fatale, Félix (1914-2002) carried on her acting career until the late 1960s. The actress, well known in Mexico and France for films such as Emilio Fernandez's Enamorada (1946), Jean Renoir's French Cancan (1955), and Luis Buñuel's Fever Rises in El Pao (1959), was a true reptile lover.

(see NE 10 A68, NE 43 A75)

双喀迈拉手镯

1957 年
卡地亚巴黎
黄金，白金
明亮式切割，长阶梯形切割和单面切割钻石
10 颗椭圆形切面红宝石
6.5 × 8.0 × 1.40 厘米

利用旋转装置打开手镯。

223

BT 62 A57

Twin *chimera*-head bangle

Cartier Paris, 1957
Yellow gold, white gold
Brilliant-, baguette- and single-cut
diamonds
Ten oval faceted rubies
6.5 × 8.0 × 1.40 cm

Swivel system allows the bracelet to
open.

224

CL 29 A57

棕榈叶胸针（一对）

1957 年
卡地亚巴黎
铂金，白金
梨形、马眼形的明亮式切割和单面切割钻石
两颗带截角长方形蓝宝石（分别重 20.35 克拉
和 19.30 克拉）
每枚胸针　4.80 × 4.20 × 0.70 厘米

这两枚胸针制作于1957年，当时
它们连同第三枚胸针一起被镶嵌在一
条钻石项链上。1958年，这两枚胸针被
分别售出，而位于中央的胸针仍保留在
项链上。

224

CL 29 A57

Pair of *Palm* clip brooches

Cartier Paris, 1957
Platinum, white gold
Pear-shaped, marquise-, brilliant- and
single-cut diamonds
Two rectangular sapphires with cut corners
(20.35 and 19.30 carats)
Each brooch 4.80 × 4.20 × 0.70 cm

When these brooches were made
in 1957, they were attached—along
with a third brooch—to a diamond
necklace. The following year they
were sold separately, while the central
brooch remained on the necklace.

225

CL 34 A57

棕榈树胸针

1957 年
卡地亚巴黎，特别订制
铂金，白金
明亮式和长阶梯形切割钻石
7 颗枕形缅甸红宝石（总重 23.10 克拉）
11.50 × 7.00 × 2.75 厘米

225

CL 34 A57

Palm-tree clip brooch

Cartier Paris, special order, 1957
Platinum, white gold
Brilliant- and baguette-cut diamonds
Seven cushion-shaped Burmese rubies,
weighing 23.10 carats in total
11.50 × 7.00 × 2.75 cm

226

CL 140 A57

虎形胸针

1957 年
卡地亚巴黎
金
单面切割和明亮式切割钻石（颜色多样，从黄色
到近乎无色）
马眼形祖母绿（眼睛）
花式切割黑玛瑙（虎纹）
7.00 × 4.50 × 1.50 厘米

售予芭芭拉·赫顿（参见RG 30
A34）。

此款珠宝为可活动链节型。

226

CL 140 A57

Tiger clip brooch

Cartier Paris, 1957
Gold
Single- and brilliant-cut diamonds ranging
from fancy intense yellow to near colorless
Marquise-shaped emeralds (eyes)
Fancy-shaped onyxes (stripes)
7.00 × 4.50 × 1.50 cm

This piece is articulated.

Sold to Barbara Hutton

(see RG 30 A34)

猎豹棘爪式别针

1957 年
卡地亚巴黎
铂金, 白金
明亮式切割和单面切割钻石
凸圆形蓝宝石, 马眼形切割祖母绿, 绿色石榴石,
黑玛瑙
9.7 × 3.6 × 1.7 厘米

猎豹手镯

1958 年
卡地亚巴黎
铂金, 白金
明亮式切割和单面切割钻石
凸圆形蓝宝石, 马眼形切割祖母绿, 绿色石榴石,
黑玛瑙
5.85 × 7.15 × 1.75 厘米

猎豹胸针

1958 年
卡地亚巴黎
铂金, 白金
明亮式切割和单面切割钻石
凸圆形蓝宝石, 马眼形切割祖母绿, 绿色石榴石,
玛瑙
5.35 × 2.65 × 2.0 厘米

来源于阿迦·汗亲王。

英印混血的尼娜·戴尔
(1930～1965年)是风华绝代的时装模
特。她的父亲是锡兰的一名茶叶种植
商。在与第一任丈夫亨尼克·泰森·宝
恩美采男爵离婚后, 她嫁给了阿迦·汗
亲王, 这段婚姻又于5年后的1962年结
束。1965年, 尼娜·戴尔凄惨地离开人
世。她生前的物品于1969年进行拍卖。
她那整柜的豹皮服装和豹形的珠宝,
证明了她对于这一动物造型的极度热
爱。她收藏的卡地亚猎豹系列珠宝还
包括另一条手链和一枚戒指。

227

JS 02 A57-58

Panther cliquet pin

Cartier Paris, 1957
Platinum, white gold
Brilliant- and single-cut diamonds
Sapphire cabochons, marquise-cut
emeralds, green garnets, onyx
9.7 × 3.6 × 1.7 cm

Panther bangle

Cartier Paris, 1958
Platinum, white gold
Brilliant- and single-cut diamonds
Sapphire cabochons, marquise-cut
emeralds, green garnets, onyx
5.85 × 7.15 × 1.75 cm

Panther clip brooch

Cartier Paris, 1958
Platinum, white gold
Brilliant- and single-cut diamonds
Sapphire cabochons, marquise-cut
emeralds, green garnets, onyx
5.35 × 2.65 × 2.0 cm

Provenance : Princess Sadruddin Aga
Khan

Of Anglo-Indian origin, the extremely beautiful fashion model Nina Dyer (1930–1965) was the daughter of a tea planter in Ceylon. After a first marriage to Baron Heinrich Thyssen-Bornemisza, she wed Sadruddin Aga Khan, whom she divorced five years later, in 1962. After her tragic death in 1965, her property was auctioned in 1969. Her possessions included a substantial wardrobe of panther-skin clothing and jewellery in the same pattern, testifying to her passion for this cat. Her collection of Cartierpanthers included another bracelet and a ring.

228

CL 55 A59

蓝色玫瑰胸针

1959 年
卡地亚巴黎为卡地亚纽约制作
铂金，白金
明亮式和长阶梯形切割钻石
圆柱形切割切面蓝宝石
4.54 × 4.03 × 2.60 厘米

228

CL 55 A59

Blue rose clip brooch

Cartier Paris for New York, 1959
Platinum, white gold
Brilliant- and baguette-cut diamonds
Faceted and calibré-cut sapphires
4.54 × 4.03 × 2.60 cm

229

CL 245 A62

龟形胸针

1962 年
卡地亚巴黎
线刻金，铂金
明亮式切割钻石
一颗凸圆形星彩蓝宝石（重 93.14 克拉），一颗
枕形蓝宝石
凸圆形绿松石
5.14 × 4.06 × 2.0 厘米

229

CL 245 A62

Tortoise clip brooch

Cartier Paris, 1962
Chased gold, platinum
Brilliant-cut diamonds
One 93.14 carat star sapphire cabochon,
one cushion-shaped sapphire
Turquoise cabochons
5.14 × 4.06 × 2.0 cm

230

CL 194 A62

花朵胸针

1962 年卡地亚巴黎, 特别订制
铂金, 白金
明亮式和长阶梯形切割钻石
一颗八角形祖母绿式切割黄色蓝宝石（重 12.49
克拉）
5.6 × 4.8 × 1.6 厘米

230

CL 194 A62

Flower clip brooch

Cartier Paris, special order, 1962
Platinum, white gold
Brilliant- and baguette-cut diamonds
One octagonal emerald-cut 12.49-carat
yellow sapphire
5.6 × 4.8 × 1.6 cm

手链

1963 年
卡地亚伦敦，特别订制
铂金
明亮式切割和长阶梯形切割钻石
80 颗椭圆形切面红宝石（总重 53.60 克拉）
长度　17.30 厘米

　　为制作这条手链，卡地亚伦敦将
一名印度王妃手链上的宝石进行了重
新组合。

231

BT 96 A63

Bracelet

Cartier London, special order, 1963
Platinum
Brilliant- and baguette-cut diamonds
Eighty oval faceted rubies, weighing 53.60
carats in total
Length 17.30 cm

For this piece, Cartier London reset
the stones from a bracelet belonging
to an Indian maharani.

232

RG 02 A64

小圆球戒指

1964 年
卡地亚巴黎
圆模雕刻装饰金, 铂金
明亮式切割钻石
红宝石珠
3.5 × 3.1 × 2.3 厘米

此款戒指由卡地亚巴黎于1935年制造。从那时起, 便成为一款经典之作。

232

RG 02 A64

Boule ring

Cartier Paris, 1964
Gadrooned gold, platinum
Brilliant-cut diamonds
Ruby beads
3.5 × 3.1 × 2.3 cm

This model was created by Cartier Paris in 1935. Since then, it has become a classic.

233

CL 164 A65

玫瑰花蕾胸针

1965 年
卡地亚巴黎
黄金, 玫瑰金
明亮式切割花式黄钻石
明亮式切割和单面切割钻石
圆形切面祖母绿
祖母绿间镶嵌绿色珐琅饰物
6.9 × 3.70 × 2.85 厘米

233

CL 164 A65

Rosebud clip brooch

Cartier Paris, 1965
Yellow gold, pink gold
Brilliant-cut fancy intense yellow diamonds
Brilliant- and single-cut diamonds
Round faceted emeralds
Green enamel studs between the emeralds
6.9 × 3.70 × 2.85 cm

234
RG 17 A67

"你和我" 戒指
（交叠戒指）

1967 年
卡地亚巴黎
圆模雕刻装饰金，铂金
明亮式切割钻石
一颗圆锥形凸圆形蓝宝石（重 3.56 克拉），
圆形切面蓝宝石
凸圆形绿松石
3.0 × 2.50 厘米

234
RG 17 A67

Toi et moi ring
(Crossover ring)

Cartier Paris, 1967
Gadrooned gold, platinum
Brilliant-cut diamonds
One sugar-loaf 3.56 carat sapphire
cabochon, round faceted sapphires
Turquoise cabochons
3.0 × 2.50 cm

235
BT 86 A67

虎形手镯

1967 年
卡地亚巴黎
金
明亮式切割和单面切割黄色钻石
两颗梨形祖母绿（眼睛）
花式切割黑玛瑙（虎纹和虎鼻）
7.0 × 6.05 × 3.0 厘米

老虎头部可旋转。

235
BT 86 A67

Tiger bracelet

Cartier Paris, 1967
Gold
Brilliant- and single-cut fancy intense
yellow diamonds
Two pear-shaped emeralds (eyes)
Fancy-shaped onyxes (stripes and nose)
7.0 × 6.05 × 3.0 cm

The head of the animal swivels.

236

NE 10 A68

蛇形项链

1968 年
卡地亚巴黎，特别订制
铂金，白金和黄金
2473 颗明亮式和长阶梯形切割钻石（总重
178.21 克拉）
两颗梨形祖母绿（眼睛）
绿色、红色、黑色珐琅
长度　57.0 厘米

为玛莉亚·菲利克斯特别订制（参
见EB 25 A57, NE 43 A75）。

236

NE 10 A68

Snake necklace

Cartier Paris, special order, 1968
Platinum, white gold and yellow gold
2 473 brilliant- and baguette-cut diamonds,
weighing 178.21 carats in total
Two pear-shaped emeralds (eyes)
Green, red and black enamel
Length 57.0 cm

Made to special order for María Félix.

(see EB 25 A57, NE 43 A75)

237
BT 09 A68

手镯

1968 年
卡地亚巴黎，特别订制
铂金，白金
马眼形和明亮式切割钻石
雕花刻槽式珊瑚
直径　8.05 厘米

237
BT 09 A68

Bangle

Cartier Paris, special order 1968
Platinum, white gold
Marquise- and brilliant-cut diamonds
Carved and fluted coral
Diameter 8.05 cm

238

BT 115 A69

海豚手镯

1969 年
卡地亚巴黎，特别订制
雕刻金
1028 颗明亮式切割钻石（总重 33.73 克拉）
2 颗梨形祖母绿（眼睛）
9.0 × 8.05 × 5.7 厘米

238

BT 115 A69

Dolphin bangle

Cartier Paris, special order, 1969
Engraved gold
1,028 brilliant-cut diamonds, weighing
33.73 carats in total
Two pear-shaped emeralds (eyes)
9.0 × 8.05 × 5.7 cm

239

CL 182 A69

龟形胸针

1969 年
卡地亚巴黎，特别订制
线刻绞纹金
明亮式切割和花式切割黄钻石
两颗凸圆形蓝宝石
3.86 × 2.68 厘米

239

CL 182 A69

Tortoise clip brooch

Cartier Paris, special order, 1969
Chased and twisted gold
Brilliant-cut fancy intense yellow diamonds
Two sapphire cabochons
3.86 × 2.68 cm

240

CL 151 A69

鹦鹉胸针

1969 年
卡地亚巴黎
金，铂金
明亮式切割和花式切割黄钻石
明亮式切割和长阶梯形切割钻石
一颗椭圆形切面祖母绿
雕花珊瑚
8.5 × 2.8 厘米

240

CL 151 A69

Parrot clip brooch

Cartier Paris, 1969
Gold, platinum
Brilliant-cut fancy intense yellow diamonds
Brilliant- and baguette-cut diamonds
One oval faceted emerald
Carved coral
8.5 × 2.8 cm

241

CL 215 A69

瓢虫胸针

1969 年
卡地亚巴黎
白金，铂金
一颗长阶梯形切割钻石，一颗半月形钻石，明亮
式切割钻石
镶钻珊瑚
黑色漆面
3.00 × 2.00 × 1.00 厘米

241

CL 215 A69

Ladybird clip brooch

Cartier Paris, 1969
White gold, platinum
One baguette-cut diamond, one half-moon
diamond, brilliant-cut diamonds
Coral studded with collet-set diamonds
Black lacquer
3.00 × 2.00 × 1.00 cm

242

BT 23 A72

蜥蜴手镯

1972 年
卡地亚巴黎
金
明亮式切割和花式切割黄钻石
明亮式切割和梨形钻石
圆形切面祖母绿
圆形切面蓝宝石
2 颗凸圆形红宝石（眼睛）
6.02 × 6.08 × 6.03 厘米

来源于若瑟蒂·狄（参见CL 64
A45, EG 26 A52, CO 13 A52）。

242

BT 23 A72

Lizard bangle

Cartier Paris, 1972
Gold
Brilliant-cut fancy intense yellow diamonds
Brilliant-cut and pear-shaped diamonds
Round faceted emeralds
Round faceted sapphires
Two ruby cabochons (eyes)
6.02 × 6.08 × 6.03 cm

Provenance: Josette Day
(see CL 64 A45, EG 26 A52, CO 13
A52)

243

NE 43 A75

鳄鱼项链

1975 年
卡地亚巴黎，特别订制
金
1023 颗明亮式切割花式切割黄钻石（总重
60.02 克拉）
2 颗马眼凸圆形祖母绿（眼睛）
1060 颗祖母绿（总重 66.86 克拉）
2 颗凸圆形红宝石（眼睛）
右侧鳄鱼长　30.0 厘米
左侧鳄鱼长　27.30 厘米

为玛莉亚·菲利克斯特别订制（参见EB 25 A57, NE 10 A68）。

243

NE 43 A75

Crocodile necklace

Cartier Paris, special order, 1975
Gold
1 023 brilliant-cut fancy intense yellow
diamonds, weighing 60.02 carats in total
Two navette-shaped emerald cabochons
(eyes)
1 060 emeralds, weighing 66.86 carats in
total
Two ruby cabochons (eyes)
Length: 30.0 cm; length: 27.30 cm

Made to special order for María Félix.
(see EB 25 A57, NE 10 A68)

卡地亚作品中的中国元素

Chinese Elements

卡地亚 20 世纪 20 ～ 30 年代的经典作品中，有相当一部分深受中国风格影响，反映出卡地亚设计师对中国艺术的迷恋。20 世纪以前，大部分没有到过中国的欧洲人通过外销艺术品认识中国，作为中国文化载体的丝绸、织绣、陶瓷、漆器、书法、绘画、玉器、家具、园林等所传达出的儒雅温和、恬静自由、富于联想、崇尚自然的理念，给欧洲装饰艺术带来创意新灵感。特别是 17 ～ 18 世纪大量中国艺术品输入欧洲，这些艺术品浓缩了中国古老文化的精髓及数千年的文化底蕴，具有永恒、智慧、神秘的魅力。充满浓郁东方情调的形式和纹饰，打破透视、比例及对称原则，激发了欧洲皇室、贵族、学者、艺术家、收藏家对中国艺术品的极大热情和收藏欲望，从收藏到仿制，最终促成中国风格装饰在欧洲盛行。

追求时尚在于理智而熟练的驾驭。卡地亚设计师不但具备敏锐捕捉社会流行趋势的能力，而且能够充满智慧的运用和驾驭流行元素。卡地亚的这些充满中国韵味的作品充分反映出设计师从中国艺术中寻找设计灵感，力求表现对中国文化的解读和诠释。作品最突出的是受中国玉雕、漆器和丝织品纹饰的影响，呈现出两个显著特点：一是借用中国艺术品直接植入新作品的设计中，对所植入的艺术品尽可能的保持原貌，并在此基础上进行二次创作。如：翡翠坠饰（参见 PE 15 A21），取材于一件中国清代荷花纹翡翠佩；龙形胸针（参见 CL 80 A24），是根据中国清代翡翠龙形带钩重新设计；烟灰缸（参见 SA 04 A35），原本是中国文房用具白玉莲花笔洗；香精瓶（参见 FK 06 A25, FK 11 A25, FK 05 A26, FK 19 A26）和打火机（参见 LR 22 A39）则是将鼻烟壶稍作加工即改变了它的用途。这些作品经过巧妙构思，新颖创意，并根据作品材质、形式，佐以符合欧洲审美趣味的珠宝装饰，不但散发着浓浓的中国韵味，而且赋予了新的意义。二是从绘画、陶瓷、漆器、丝织品等中国艺术品中择取文化元素融入设计中，经过二次创作扩展了这些文化元素的内涵。如：中国的太极两仪图形被用在链坠挂饰（参见 NE 19 A19）中；"寿"字和弥勒佛耳坠（参见 EG 24 A26, EG 11 A28），不仅延续长寿和吉祥的寓意，同时也是祈福的护身符。此外，龙、凤、犬、瑞兽等经过设计师极具创意的设计赋予了新的生命。碧玉和珊瑚的大胆搭配，呈现出典型的中国色彩；与中国嵌螺钿工艺结合，使其具有中国纹饰的特点。晶莹华丽，美轮美奂，释放出迷人魅力。

卡地亚这些具有中国风格的作品，无论是设计中直接植入中国艺术品，还是择取中国文化元素融入设计中，都显示了卡地亚作品所拥有的文化内涵和典雅、高贵气质，也是卡地亚作品的恒久魅力所在。

故宫博物院　宋海洋

Cartier was profoundly influenced by Chinese style throughout the 1920s and 1930s, reflecting a resounding dedication to Chinese art. Before the 20th century, most Europeans knew about China through the exported Chinese artifacts, including the silk embroideries, ceramics, lacquerware, calligraphies, paintings, jades, furniture and garden models which were exported to Europe during the 17th and 18th centuries. The artifacts conveyed Chinese concepts of grace, harmony, freedom and love for nature, inspired European decorative arts with Chinese wisdom and mysterious charm. The forms and ornamentation with their profound oriental exoticism shattered previous conceptions of perspective, proportion and symmetry, which inculcated a great interest and desire to collect Chinese artifacts amongst the European Royal Courts, aristocrats, scholars, artists and connoiseurs. They collected and imitated Chinese artworks, making Chinese-style decorations popular throughout Europe.

The leadership of trend lies in the rational and skillful mastery of design. Cartier designers captured social trends as while intelligently using and controlling popular elements. Cartier creations endorsed a Chinese appeal, reflecting the efforts of the designers in interpreting Chinese culture. These creations were most influenced by jade carvings, lacquerware, and silk ornamentation. This influence has two features. One is to directly borrow Chinese artworks, preserve the original styles, then begin new creations based on them. For example, the jade pendant (see PE 15 A21) taken from a piece of Chinese lotus-shaped jade of the Qing Dynasty; a dragon brooch (see CL 80 A24) redesigned on the Chinese dragon-shaped jade belt hook of the Qing Dynasty; an ashtray (see SA 04 A35) from a Chinese lotus-shaped white jade ink stone; perfume bottles (see FK 06 A25, FK 11 A25, FK 05 A26, FK 19 A26); and a lighter (see LR 22 A39) designed with a little modification on Chinese snuff bottles. They were carefully and innovatively redesigned to meet European aesthetic tastes in materials and forms, while still keeping the original Chinese styles and meanings. The other way is to integrate Chinese elements of paintings, porcelain, lacquerware and silk into the design, hence expanded the definitions of cultural elements through their creations. For example, China's *Yin* and *Yang* in Taiji is used in a pendant (see NE 19 A19); a pair of ear-pendants in the shape of Chinese character "Shou" and Buddhas (see EG 24 A26, EG 11 A28), respectively meaning longevity and auspiciousness, which symbolize blessings. The patterns of the dragon, phoenix, dog, and propitious animal were endowed with new lives through their creative design. The collocation of jade and coral presents the typical Chinese style; a mother-of-pearl inlay is similar to the traditional Chinese handicraft. These combinations presents the charm of glitter, translucence, glory, and beauty.

Cartier creations, whether directly embedded with the artifacts or integrated with selected elements, reflect the elegance and nobility giving an everlasting charm!

Song Haiyang
The Palace Museum

糖果盒

1912 年
卡地亚巴黎
铂金，金
雕花翡翠
玫瑰式切割钻石，两颗凸椭圆形红宝石，两颗方
形切割切面红宝石
5.33 × 2.33 厘米

售予托比伯爵夫人。

托比伯爵夫人是俄国大公米歇
尔·米哈伊诺维奇的平民妻子。

244

BS 05 A12

Bonbonniere

Cartier Paris, 1912
Platinum, gold
Carved jade
Rose-cut diamonds, two oval-shaped ruby
cabochons, two square-cut and faceted
rubies
5.33 × 2.33 cm

Sold to the Countess of Torby,
morganatic wife of Grand Duke
Michael Michailovich of Russia.

245

NE 19 A19

阴阳链坠

1919 年
卡地亚巴黎
铂金，金
圆形旧式切割、单面切割和玫瑰式切割钻石
两颗钻孔祖母绿珠
凸圆形切割红宝石、圆柱形切割红宝石
一颗祖母绿，四颗红宝石和四颗面弧形黑玛瑙（链
绳）
黑玛瑙
黑色珐琅（搭扣）
长度　31.5 厘米

245

NE 19 A19

Yin-Yang pendant

Cartier Paris, 1919
Platinum, gold
Round old-, single- and rose-cut diamonds
Two drilled emerald beads
Calibré-cut rubies and collet-set ruby
cabochons
One emerald, four rubies and four buff-top
onyxes on the loop
Onyx
Black enamel on the clasp
Length 31.5 cm

246

旋转时标时钟

1920 年
卡地亚巴黎
金，镀金金属
中央为描绘中国景物的螺钿漆画（参见 CDS 06 A29）
黑色和白色珐琅
硬橡胶（背后和边缘）
直径 9.2 厘米

外壳背后镌刻以下文字：（凯特·史密斯从事广播播音工作五周年志喜，威廉·S·帕里）。

以12个白色珐琅球形旋涡和金色罗马数字刻度指示小时，依次沿表盘旋转，指示分钟数。

此钟为8日储存圆形机芯，镀金，2个调校项目，12颗宝石轴承，瑞士杠杆式擒纵结构，双金属平衡摆轮，宝玑摆轮游丝。

威廉·S·帕里是美国三大广播公司之一的哥伦比亚广播公司总裁。他的妻子芭比·佩利（全名芭芭拉·辜辛·佩利）是纽约上流社会成员，被誉为全球最优雅的女性之一。

246

Clock with rotating hours

Cartier Paris, 1920
Gold, gilded metal
The center of laque burgauté depicting a
Chinese scene (see CDS 06 A29)
Black and white enamel
Ebonite (back and edge)
Diameter 9.2 cm

Engraved on back of case: To Kate Smith on her fifth anniversary in radio broadcasting in appreciation from William S. Paley

Hours indicated by 12 bulb-shaped cartouches of white enamel with gold Roman numerals in reserve, which successively rotate across the dial and point to the minutes.

Round 8-day movement, gold-plated, 2 adjustments, 12 jewels, Swiss lever escapement, bimetallic balance, Breguet balance spring.

William S. Paley was head of the Columbia Broadcasting System (CBS), one of the 3 major US network. His wife Babe Paley (Barbara Cushing Paley), was a member of New York's high society, and one of the most elegant women in the world.

吊坠

1921 年
卡地亚巴黎
铂金
雕花翡翠
凸圆形切割和圆柱形切割红宝石
玫瑰式切割钻石
8.50 × 3.20 厘米

这块翡翠双面均雕刻荷叶和荷花。
这个吊坠取材于清代（1644～1911年）
的同类饰物。

247

PE 15 A21

Pendant

Cartier Paris, 1921
Platinum
Carved jade
Cabochon- and calibré-cut rubies
Rose-cut diamonds
8.50 × 3.20 cm

The jade plaque is carved on both
sides with lotus leaves and flowers.
This pendant is based on similar jewels
from the Qing Dynasty (1644-1911).

248

NE 03 A22

吊坠

1922 年
卡地亚巴黎
铂金
圆形旧式切割、单面切割和玫瑰式切割钻石
一颗祖母绿（眼睛）
天然珍珠
珊瑚，黑玛瑙珠和圆环
吊坠长度　18.50 厘米

售予凯德尔斯顿的寇松侯爵夫人。

这款吊坠的灵感取材于中国式护身符，其佩戴习俗可一直追溯到商代（公元前2000年）。

格莉斯·埃尔维拉·亨兹（1879～1958年），美国外交家约瑟夫·蒙罗·亨兹之女，初嫁与阿尔弗雷德·杜甘，后于1917年成为乔治·纳撒尼尔——凯德尔斯顿的寇松侯爵（1859～1925年）的第二任妻子。寇松侯爵曾于1898年到1905年期间出任印度总督。1955年，寇松侯爵夫人出版了自己的《回忆录》。

248

NE 03 A22

Pendant

Cartier Paris, 1922
Platinum
Round old-, single- and rose-cut diamonds
One emerald (eye)
Natural pearls
Coral, onyx beads and roundels
Length18.50cm

This pendant was inspired by certain Chinese pendant-amulets, traditionally worn since the Shang Dynasty (second millennium B.C.E.).

Sold to the Marchioness Curzon of Kedleston

Grace Elvina Hinds (1879–1958), daughter of American diplomat Joseph Monroe Hinds, first married Alfred Duggan and in 1917 became the second wife of George Nathaniel, Marquess Curzon of Kedleston (1859–1925), Viceroy of Indian from 1898 to 1905. In 1955 she published her Reminiscences.

249

CL 258 A22

胸针

1922 年
卡地亚巴黎
铂金
圆形旧式切割及单面切割钻石
一颗祖母绿（重 19.45 克拉），一颗钻孔祖母绿珠，
两颗凸圆形祖母绿
珊瑚
黑玛瑙
13.40 × 3.35 厘米

售予威廉·K·范德比尔特夫人
（参见CL 92 A05, CL 06 A25, CL 244
A25）。

249

CL 258 A22

Brooch

Cartier Paris, 1922
Platinum
Round old- and single-cut diamonds
One pear-shaped 19.45 carat emerald, one
drilled emerald bead, emerald cabochons
Coral
Onyx
13.40 × 3.35 cm

Sold to Mrs. William K. Vanderbilt
    (see CL 92 A05, CL 06 A25, CL 244
A25)

250

TB 07 A22

座式雪茄盒和烟盒

1922 年
卡地亚巴黎
银，金
螺钿
红漆
珊瑚，黑玛瑙
檀木
23.80 × 17.80 × 7.93 厘米

盒盖嵌染色螺钿中国风景图案。

内设两个隔层，包括一个香烟盒，一个雪茄盒。另有两个火柴盒，其中一个带盒盖和划火面。

250

TB 07 A22

Cigar and cigarette table box

Cartier Paris, 1922
Silver, gold
Mother-of-pearl
Red lacquer
Coral, onyx
Ebony
23.80 × 17.80 × 7.93 cm

The lid inset with a Mosaic in dyed mother-of-pearl depicting a Chinese scene.

The interior fitted with two compartments, one for cigarettes and one for cigars, and two match containers, one with cover and striking surface.

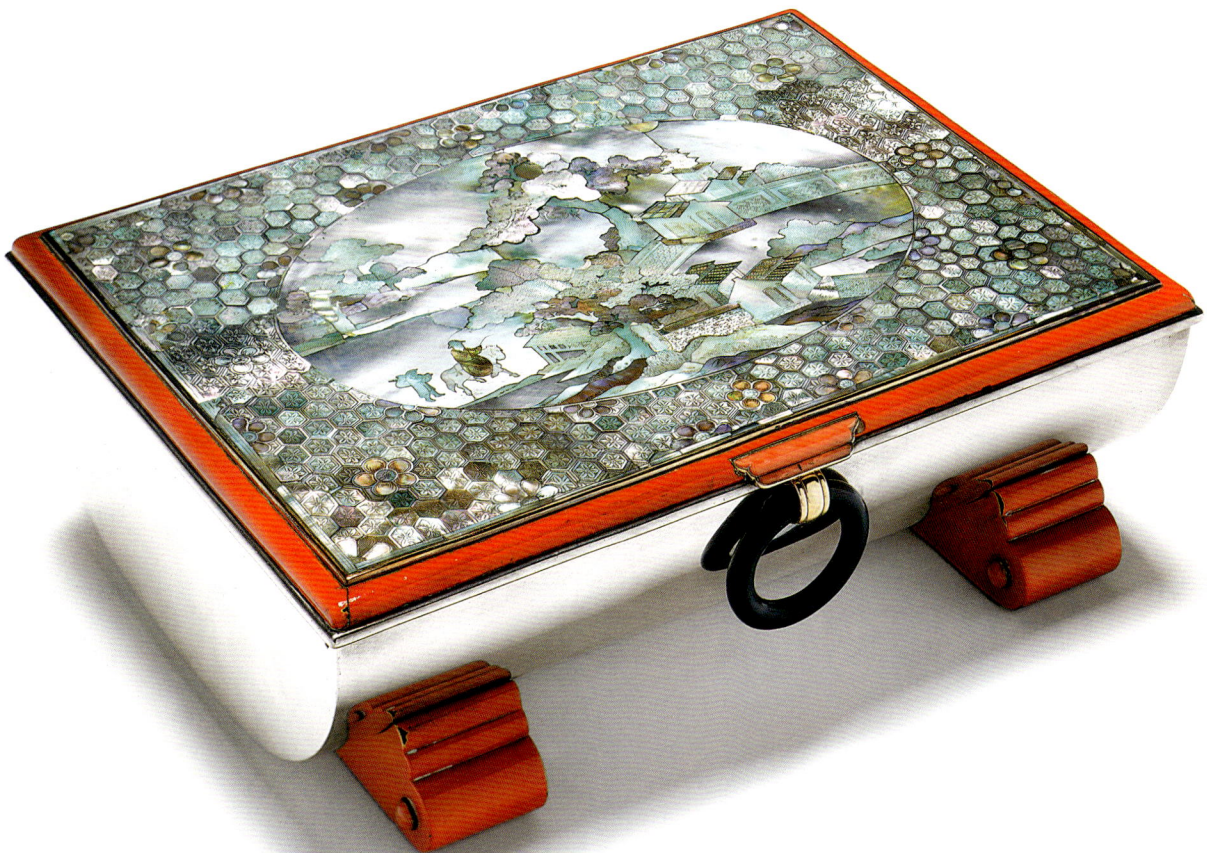

251

CL 260 A23

喀迈拉棘爪式扣针

1923 年
卡地亚巴黎
铂金，金
圆形旧式切割及单面切割钻石
一颗刻槽式祖母绿珠
凸圆形祖母绿和面弧形祖母绿
一颗天然珍珠
雕花刻槽式珊瑚
凸圆形黑玛瑙
黑色珐琅
13.0 × 1.8 厘米

售予路易·卡地亚。

251

CL 260 A23

Chimera cliquet pin

Cartier Paris, 1923
Platinum, gold
Round old- and single-cut diamonds
One fluted emerald bead
Emerald cabochons and buff-top emeralds
One natural pearl
Carved and fluted coral
Onyx cabochons
Black enamel
13.0 × 1.8 cm

Sold to Louis Cartier

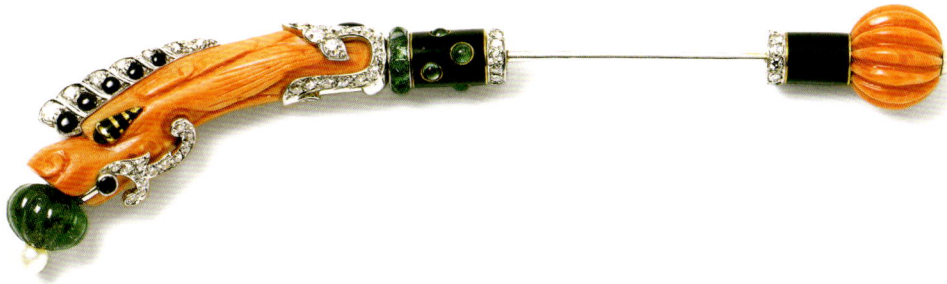

252

CM 09 A23

门廊魅幻时钟

1923 年
卡地亚巴黎
铂金，金
无色水晶表盘、立柱和福神像
玫瑰式切割钻石
凸圆形珊瑚，黑玛瑙
黑色珐琅
35.0 × 23.0 × 13.0 厘米

售予 H·F·麦考密克夫人（加娜·瓦斯卡）。

此钟为8日储存方形机芯，双发条盒，镀金，13颗宝石轴承，双金属平衡摆轮，宝玑摆轮游丝。无色水晶传动轴，凸圆形珊瑚。可移动福神像调节机芯。上链和时间调校轴。

这款时钟最早是一套日本神道教"庙门"形时钟中的一个。这套时钟共有六款，每一款都各不相同，由卡地亚在1923～1925年之间创作。

加娜·瓦斯卡（1887～1984年）是一名波兰裔歌剧演员，嫁给有着"全球身价最高的黄金单身汉"之称的亚历山大·史密斯·科克伦，后改嫁农业机械制造厂麦考密克继承人哈罗德·福勒·麦考密克（参见JA 26 A30，BT 109 A28）。

252

CM 09 A23

Large *Portique* mystery clock

Cartier Paris, 1923
Platinum, gold
Rock crystal dial, columns and Biliken figure
Rose-cut diamonds
Coral cabochons, onyx
Black enamel
35.0 × 23.0 × 13.0 cm

Square 8-day double-barrel movement, gold-plated, 13 jewels, bimetallic balance, Breguet balance spring. Transmission axle in rock-crystal crossbar masked by coral cabochon. Billiken figure removable to provide access to the movement. Arbor for winding movement and setting hands.

This clock was the first in a series of 6 in the form of a Shinto "shrine gate" (portique), all different and all made by Cartier between 1923 and 1925.

Sold to Mrs. H.F. McCormick (Ganna Walska) of Polish origin, the opera singer Ganna Walska (1887-1984) was married to Alexander Smith Cochran (dubbed "the richest bachelor in the world") and later Harold Fowler McCormick, heir to the McCormick farm machinery fortune.

(see JA 26 A30, BT 109 A28)

Fait par Cartier   Paris-Londres-New-York

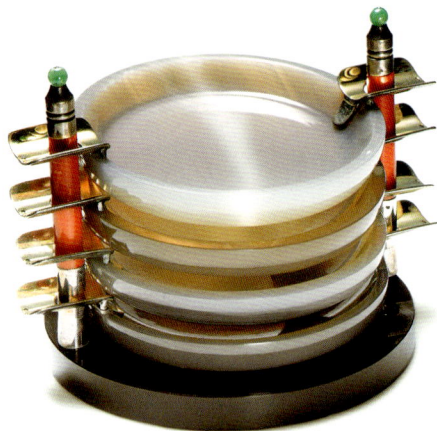

253

SA 03 C23

烟灰缸

约 1923 年
卡地亚巴黎
银，银镀金
玛瑙（烟灰缸）
翡翠石珠
珊瑚（支柱）
黑玛瑙（底座）
7.0 × 6.0 × 9.50 厘米

253

SA 03 C23

Ashtrays

Cartier Paris, c. 1923
Silver, silver-gilt
Agate (ashtrays)
Jade beads
Coral (columns)
Onyx (base)
7.0 × 6.0 × 9.50 cm

254

CL 80 A24

龙形胸针

1924 年
卡地亚巴黎，特别订制
金，铂金
圆形旧式切割和单面切割钻石
凸圆形蓝宝石
翡翠
黑色珐琅
9.45 × 1.85 厘米

这枚胸针由客户所提供的18～19
世纪的翡翠带钩制成。

254

CL 80 A24

Dragon brooch

Cartier Paris, special order, 1924
Gold, platinum
Round old- and single-cut diamonds
Sapphire cabochons
Carved jade
Black enamel
9.45 × 1.85 cm

This brooch was made from a carved
jade Chinese belt clasp of the eighteenth or nineteenth century, supplied
by the client.

255

VC 67 A24

化妆盒

1924 年
卡地亚巴黎
金，铂金
螺钿，绿松石，珊瑚
青金石，孔雀石，蓝孔雀石，玛瑙
玫瑰式切割钻石
黑玛瑙，黑色珐琅
木
11 × 6.50 × 1.70 厘米

内设一面化妆镜和一个弹簧隔层。

化妆盒正面描绘了三个汉族人中的两个，旁边是代表长寿的鹳鸟和绵羊，背面是一个道教的神仙和动物（中国，19世纪）。顶部和底部均带有"寿"字纹（参见汉斯·纳德霍夫《卡地亚——非同凡响的珠宝商》，1984年版，彩图34）。

255

VC 67 A24

Vanity case

Cartier Paris, 1924
Gold, platinum
Mother-of-pearl, turquoise, coral
Lapis lazuli, malachite, azurmalachite, agate
Rose-cut diamonds
Onyx, black enamel
Wood
11 × 6.50 × 1.70 cm

The interior fitted with a mirror and one compartment with spring holder.

The recto of the case depicting two of the three Han heroes accompanied by the stork of long life and sheep, the verso with a Taoist immortal and animals (Chinese, nineteenth century). "Shou" emblem of long life at top and bottom (see Hans Nadelhoffer, Cartier, Jewellers Extraordinary, Thames and Hudson, 1984; colour plate 34).

256

CS 11 A25

带飞返式指针鲤鱼时钟

1925 年
卡地亚巴黎
铂金，金
玉（鲤鱼）
黑曜石（基座）
无色水晶
螺钿，珍珠
珊瑚
凸圆形祖母绿，玫瑰式切割钻石
紫红色漆面，黑色、蓝色、红色珐琅
23.0 × 23.0 × 11.0 厘米

此钟为8日储存长方形机芯，镀金，飞返式时针，标准擒纵机构，双金属平衡摆轮，扁平摆轮游丝。由于时针不能整圈旋转，因此在到达右边的六点位置（VI）以后，就会弹回起点，因此得名为"飞返指针"（参见CM 23 A26，CM 04 A31）。

玉鲤来自18世纪的中国。从严格意义上来说，这一款时钟并不是魅幻时钟，而是一系列12款动物或小雕像时钟中的一款。

256

CS 11 A25

Carp clock with retrograde hand

Cartier Paris, 1925
Platinum, gold
Gray jade (carps)
Obsidian (base)
Rock crystal
Mother-of-pearl, pearls
Coral
Emerald cabochons, rose-cut diamonds
Mauve lacquer, black, blue and red enamel
23.0 × 23.0 × 11.0 cm

The jade carps are Chinese in origin, dating from the 18th century. Although not a mystery clock, strictly speaking, this is the third in a series of 12 clocks featuring animals or figurines.

Rectangular 8-day movement, gilded, retrograde hour hand, platform escapement, bimetallic balance, flat balance spring.

Since the hour hand cannot make a complete rotation, when it reaches the VI on the right it springs back to the start, hence the name "retrograde hand."

(see CM 23 A26, CM 04 A31)

257
VC 58 A25

凤凰粉盒及唇膏盒

1925 年
卡地亚巴黎
金，铂金
凸圆形祖母绿
象牙，螺钿
玫瑰式切割钻石
红色、黑色珐琅
12.5 × 6.38 厘米
粉盒直径 4.9 厘米

内设一面镜子和一个化妆品隔层。

257
VC 58 A25

Phoenix powder compact
with lipstick holder

Cartier Paris, 1925
Gold, platinum
Emerald cabochons
Ivory, mother-of-pearl
Rose-cut diamonds
Red and black enamel
12.5 × 6.38 cm;
4.9 cm (diameter of the powder compact)

The interior fitted with a mirror and a
powder compartment.

258
VC 54 A25

化妆盒

1925 年
卡地亚巴黎
金
凸圆形蓝宝石，凸圆形祖母绿，玫瑰式切割钻石
螺钿
红色、黑色珐琅
9.10 × 5.46 × 1.85 厘米

内设一面镜子，两个翻盖隔层。

258
VC 54 A25

Vanity case

Cartier Paris, 1925
Gold
Sapphire cabochons, emerald cabochons,
rose-cut diamonds
Mother-of-pearl
Red and black enamel
9.10 × 5.46 × 1.85 cm

The interior fitted with a mirror and
two lidded compartments.

259

FK 11 A25

香精瓶

1925 年
卡地亚巴黎
玉
金
凸圆形蓝宝石
黑色、蓝色珐琅
6.2 × 4.5 × 2.5 厘米

这个香精瓶是由一个中国古鼻烟
壶改制而成。

259

FK 11 A25

Scent bottle

Cartier Paris, 1925
Carved jade
Gold
Sapphire cabochons
Black and blue enamel
6.2 × 4.5 × 2.5 cm

This scent bottle was made from an
antique Chinese snuff box.

260

FK 06 A25

香精瓶

1925 年
卡地亚巴黎
碧玉
金
凸圆形红宝石
黑色、红色珐琅
6.87 × 3.4 × 2.4 厘米

260

FK 06 A25

Scent bottle

Cartier Paris, 1925
Carved nephrite
Gold
Cabochon rubies
Black and red enamel
6.87 × 3.4 × 2.4 cm

261

TB 11 A25

座式烟盒

1925 年
卡地亚
银，金
两幅描绘中国风景的螺钿漆画
红宝石，蓝宝石
月长石
翡翠
珊瑚
木，硬橡胶
6.50 × 19.80 × 9.30 厘米

（参见CDS 06 A29）

261

TB 11 A25

Cigarette table box

Cartier, 1925
Silver, gold
Two plaques of *laque burgauté* depicting
Chinese scenes
Rubies, sapphires
Moonstones
Carved jade
Coral
Wood, ebonite
6.50 × 19.80 × 9.30 cm

(see CDS 06 A29)

262

EG 24 A26

耳坠（一对）

1926 年
卡地亚纽约
铂金
圆形旧式切割和单面切割钻石
两个翡翠环
凸圆形珊瑚和珊瑚珠
红色珐琅字
4.9 × 2.4 厘米

珐琅图案是一个简化的"寿"字，寓意长寿。

262

EG 24 A26

Pair of ear-pendants

Cartier New York, 1926
Platinu m
Round old- and single-cut diamonds
Two jade rings
Coral cabochons and beads
Red enamel for the Chinese characters
4.9 × 2.4 cm

The enamel motif is a simplified version of the *shou* character, which means "long life." The red enameling imitates Chinese lacquer.

263

CM 23 A26

喀迈拉魅幻时钟

1926 年
卡地亚纽约
金，铂金
黄水晶（表盘）
玛瑙（喀迈拉），玉（波浪）
黑玛瑙
珊瑚
玫瑰式切割钻石
珍珠
凸圆形祖母绿
红色、黑色珐琅
17.0 × 13.80 × 7.45 厘米

此钟为8日储存长方形机芯，镀金，15颗宝石轴承，双金属平衡摆轮，宝玑摆轮游丝。传动轴外层覆雕花珊瑚，位于喀迈拉下方。时针调校和上链装置位于基座下方。

玛瑙神兽源自19世纪的中国。这款魅幻时钟是1922年到1931年间所创作的12款以动物或神话生物为主题的时钟中的第六款，灵感取材于路易十五和路易十六时期的时钟，其时通常将机芯安置在动物的背部。今天，本系列时钟与门廊魅幻时钟一同成为了最名贵的卡地亚标志性藏品（参见CS 11 A25, CM 04 A31）。

263

CM 23 A26

Chimera mystery clock

Cartier New York, 1926
Gold, platinum
Citrine (dial)
Agate (Chimera), nephrite (waves)
Onyx
Coral
Rose-cut diamonds
Pearls
Emerald cabochons
Red and black enamel
17.0 × 13.80 × 7.45 cm

Rectangular 8-day movement, gold-plated, 15 jewels, bimetallic balance, Breguet balance spring. Transmission axle masked a carved piece of coral beneath the chimeras. Hand-setting and winding mechanism underneath the base.

The agate chimera, of Chinese origin, dates from the 19th century. This mystery clock was the 6th in a series of 12 that featured animals or figurines, made between 1922 and 1931, partly inspired by Louis XV and Louis XVI clocks in which the movement was set on the back of an animal. Today they are considered, with the *Portique* Mystery clocks, the most valuable of all collectors' items with the Cartier signature.

(see CS 11 A25, CM 04 A31)

264

CDB 12 A26

佛犬时钟

1926 年
卡地亚巴黎
铂金, 金
钻石, 红宝石, 祖母绿, 蓝宝石
珍珠, 螺钿
无色水晶（平台、柱子以及佛犬）
珊瑚
青金石
硬橡胶（底座）
黑色、红色珐琅
22.0 × 18.0 × 6.0 厘米

此钟为8日储存浪琴圆形1941机芯, 镀金, 19个宝石轴承, 瑞士杠杆式擒纵结构, 双金属平衡摆轮, 宝玑摆轮游丝。

264

CDB 12 A26

Clock with *Fô Dogs*

Cartier Paris, 1926
Platinum, gold
Diamonds, rubies, emeralds, sapphires
Pearls, mother-of-pearl
Rock crystal (platform, column and Fô dogs)
Coral
Lapis lazuli
Ebonite (base)
Black and red enamel
22.0 × 18.0 × 6.0 cm

Round Longines 8-day caliber 1941 movement, gold-plated, 19 jewels, Swiss lever escapement, bimetallic balance, Breguet balance spring.

265

CDB 05 A26

座钟

1926 年
卡地亚巴黎
金，银镀金
玛瑙(底座)，黑玛瑙(柱脚)
翡翠，螺钿，珊瑚
玫瑰式切割钻石
珊瑚色、黑色珐琅
高度 13.50 厘米

此钟为8日储存圆形机芯，镀金，
瑞士杠杆式擒纵结构，双金属平衡摆
轮，宝玑摆轮游丝。

265

CDB 05 A26

Desk clock

Cartier Paris, 1926
Gold, silver-gilt
Agate (base), onyx (plinth)
Jade, mother-of-pearl, coral
Rose-cut diamonds
Coral-coloured and black enamel
Height 13.50 cm

Round 8-day movement, gold-plated,
15 jewels, Swiss lever escapement,
bimetallic balance, Breguet balance
spring.

CDB 23 A26

插屏式座钟

1926 年
卡地亚
铂金，金
白玉（钟盘）
黑玛瑙，珊瑚，螺钿
玫瑰式切割钻石
红色、黑色珐琅
32.7 × 29.5 × 12.0 厘米

　　钟盘没有玻璃，而是由有中国风景人物图案的白玉琢制而成。

　　此钟为8日储存长方形机芯，镀金，瑞士杠杆式擒纵结构，双金属平衡摆轮，宝玑摆轮游丝。传动轴被龙形装饰以及屏风下的雕花珊瑚球所掩盖。手动调节指针以调校时间，上弦位置在底座后部。

　　一件类似的藏品存于巴黎装饰艺术博物馆。

266

CDB 23 A26

Screen clock

Cartier, 1926
Platinum, gold
White jade (dial)
Onyx, coral, mother-of-pearl
Rose-cut diamonds
Red and black enamel
32.7 × 29.5 × 12.0 cm

The dial, without glass, is of white jade carved with a Chinese scene in front and a landscape in back.

Rectangular 8-day movement, gold-plated, Swiss lever escapement, bimetallic balance, Breguet balance spring. Transmission axle masked by the dragon and the carved coral ball beneath the screen. The time is set by manually turning the hands; winding is done at the back of the base.

A similar model is held by the Musée des Arts Décoratifs in Paris.

267

VC 14 A26

化妆盒

1926 年
卡地亚巴黎
金，铂金
两个镂空中式鹳嘴饰板
玫瑰式切割钻石
黑色珐琅
3.8 × 4.3 × 9.5 厘米

内部斜面镜背后配有香烟盒，翻盖化妆品隔层和唇膏架

267

VC 14 A26

Vanity case

Cartier Paris, 1926
Gold, platinum
Two pierced stork beak Chinese plaques
Rose-cut diamonds
Black enamel
3.8 × 4.3 × 9.5 cm

The interior fitted with a cigarette compartment at the back of the beveled mirror, a lidded powder compartment and lipstick holder.

268

FK 05 A26

香精瓶

1926 年
卡地亚巴黎
锈色雕花玉
金
青金石
凸圆形蓝宝石
黑色珐琅
6.4 × 4.6 × 2.5 厘米

268

FK 05 A26

Scent bottle

Cartier Paris, 1926
Rust-coloured carved jade
Gold
Lapis lazuli
Sapphire cabochons
Black enamel
6.4 × 4.6 × 2.5 cm

269

FK 19 A26

香精瓶

1926 年
卡地亚巴黎，特别订制
雕花珊瑚（19 世纪中国鼻烟盒）
金，铂金
珍珠，玫瑰式切割钻石
硬橡胶
黑色珐琅
螺钿
8.7 × 4.4 × 3.4 厘米

售予哈里森·威廉姆斯夫人，即蒙娜·俾斯麦伯爵夫人（参见DI 07 C27，DI 08 C27）。

269

FK 19 A26

Scent bottle

Cartier Paris, special order 1926
Carved coral (nineteenth century Chinese snuffbox)
Gold, platinum
Pearl, rose-cut diamonds
Ebonite
Black enamel
Mother-of-pearl
8.7 × 4.4 × 3.4 cm

Sold to Mrs. Harrison Williams, Countess Mona Bismarck
(see DI 07 C27, DI 08 C27)

270

OV 02 A26

带盖壶

1926 年
卡地亚巴黎
金
玛瑙
珊瑚
黑玛瑙
黑色珐琅
12.6 × 6.4 厘米

270

OV 02 A26

Pot with cover

Cartier Paris, 1926
Gold
Agate
Coral
Onyx
Black enamel
12.6 × 6.4 cm

271

OV 01 A26

花瓶

1926 年
卡地亚巴黎，特别订制
金
螺钿漆画
红漆（瓶肩和瓶边）
12.8 × 4 × 4 厘米

这个花瓶是詹姆斯·德·罗斯柴尔德先生（1878～1957年）所订制的圣诞礼物。这批礼物一共包括34个花瓶，这是其中一个（参见CDS 06 A29）。

271

OV 01 A26

Vase

Cartier Paris, special order, 1926
Gold
Plaques (laque burgauté)
Red lacquer (shoulders and rim)
12.8 × 4 × 4 cm

This vase was one of thirty-four commissioned by James de Rothschild (1878-1957) as Christmas presents. (see CDS 06 A29)

272

CL 47 A27

中国花瓶胸针

1927 年
卡地亚纽约
铂金，金
圆形旧式切割、单面切割及玫瑰式切割钻石
叶形雕花红宝石
雕花青金石
黑玛瑙珠
黑色珐琅
3.95 × 4.92 × 0.60 厘米

这个装满水果和花朵的花瓶图案，灵感源于自明朝以后中国瓷器和其他装饰品上常用的图案。

272

CL 47 A27

Chinese vase brooch

Cartier New York, 1927
Platinum, gold
Round old-, single- and rose-cut diamonds
Leaf-shaped engraved rubies
Engraved lapis lazuli
Onyx beads
Black enamel
3.95 × 4.92 × 0.60 cm

This vase filled with fruit and flowers was inspired by a motif commonly found on porcelain and other Chinese decorative objects since the Ming period.

273

CDB 07 A27

时钟

1927 年
卡地亚巴黎
金，铂金
雕有鸟和叶子图案的玛瑙（原型为中国 19 世纪
的笔筒）
切面黄水晶
青金石（底座和钟盘）
玫瑰式切割钻石
黑色珐琅
高度　13.5 厘米

此钟为8日储存长方形机芯，镀
金，17个宝石轴承，瑞士杠杆式擒纵结
构，双金属平衡摆轮，宝玑摆轮游丝。
机芯置于黄水晶下的镀金金属盒中
（原型为笔筒）。表壳下设有手动调校
时间和上弦机制。机芯和指针之间的
传动轴横穿半宝石。

273

CDB 07 A27

Clock

Cartier Paris, 1927
Gold, platinum
Agate carved with birds and foliage
(originally a brush holder, China, 19th
Century)
Faceted citrine
Lapis lazuli (base and dial)
Rose-cut diamonds
Black enamel
Height 13.5 cm

Rectangular 8-day movement, gold-
plated, 17 jewels, Swiss lever escapement, bimetallic balance, Breguet balance spring. The movement is placed
in a gilded metal box underneath the
citrine, inside the former brush holder.
Hand-setting and winding mechanism underneath the case. Transmission axle between movement and
dial traverses the semi-precious stone.

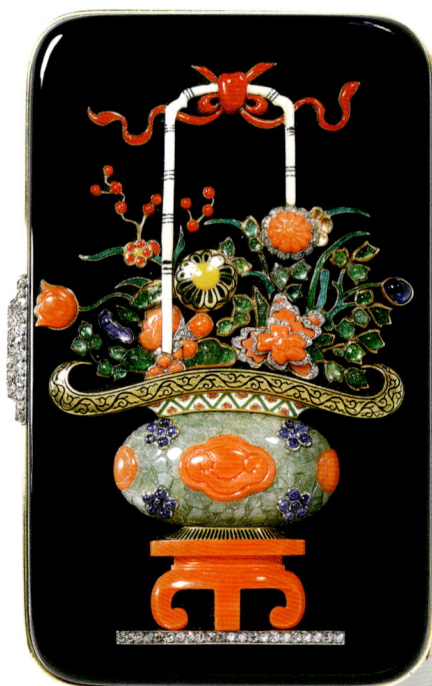

中国花瓶装饰化妆盒

1927 年
卡地亚巴黎
金，铂金
雕花祖母绿，凸圆形祖母绿，凸圆形蓝宝石
单面切割及玫瑰式切割钻石
黑玛瑙
珊瑚
红色、象牙色、黑色、绿色、黄色珐琅
9.0 × 5.9 × 2.5 厘米

内设一面镜子，两个翻盖隔层和
一个唇膏架。

Vanity case with Chinese
vase decoration

Cartier Paris, 1927
Gold, platinum
Carved emerald, emerald cabochons,
sapphire cabochons
Single- and rose-cut diamonds
Onyx
Coral
Red, ivory-coloured, black, green and
yellow enamel
9.0 × 5.9 × 2.5 cm

The interior fitted with a mirror, two
lidded compartments and a lipstick
holder.

祥龙化妆盒

1927 年
卡地亚巴黎
玫瑰金，铂金
方形和玫瑰式切割钻石
凸圆形祖母绿
圆模雕刻装饰黑玛瑙
红色、黑色、蓝色珐琅
9.2 × 5.9 × 1.5 厘米

内设一面镜子，两个化妆品隔层和一个唇膏架。

红色的底色是模仿中国漆制品。而龙的图案是来自18～19世纪中国绣花织品图案。

275

VC 69 A27

Dragon vanity case

Cartier Paris, 1927
Pink gold, platinum
Square-shaped and rose-cut diamonds
Emerald cabochons
Gadrooned onyx
Red, black and blue enamel
9.2 × 5.9 × 1.5 cm

The interior fitted with a mirror, two powder compartments and a lipstick holder.

The red ground was no doubt intended to imitate Chinese lacquer. The dragon itself is probably taken from Chinese embroidered textiles of the eighteenth and nineteenth Centuries.

276

VC 37 A27

昼夜化妆盒

1927 年
卡地亚巴黎
金, 铂金
凸圆形蓝宝石, 凸圆形红宝石, 长阶梯形及玫瑰
式切割钻石
螺钿, 红玉髓, 各种石玉
黑色珐琅
2.00 × 9.30 × 5.30 厘米

内设一面镜子, 两个翻盖隔层和一个唇膏架。

两个中国风情的螺钿饰板由弗拉迪米尔　马科斯基设计。其中一个描绘了白天的景色(红宝石代表太阳), 另一个描绘了夜晚的景色(蓝宝石代表月亮)。

弗拉迪米尔　马科斯基(1884—1966年)是长居巴黎的俄国艺术家。他以用漆绘和各种石玉来设计东方情调的艺术品见长。很多装饰艺术的经典镶嵌杰作都出自弗拉迪米尔　马科斯基之手。他的作品易于辨认——上面都刻有细小的黑色"M"字母(参见VC 58 A25, PB 21 A27)。

276

VC 37 A27

Day and night vanity case

Cartier Paris, 1927
Gold, platinum
Sapphire cabochon, ruby cabochon,
baguette- and rose-cut diamonds
Mother-of-pearl, cornelian, various hard
stones, nephrite
Black enamel
2.00 × 9.30 × 5.30 cm

The interior fitted with a mirror, two lidded compartments and a lipstick holder.

Two mother-of-pearl plaques of Chinese inspiration attributed to Vladimir Makowsky, one representing a diurnal landscape (ruby-set sun) the other a nocturnal landscape (sapphire-set moon).

Vladimir Makowsky (1884-1966) was a Russian artist living in Paris; he was specialized in creations of objects with lacquer and hard stones inspired from oriental models. Vladimir Makowsky produced some of the most exceptional inlay work of the Art Deco period. His panels can often be identified by his signature - a small black " M " in script.

(see VC 58 A25, PB 21 A27)

277

PB 21 A27

粉盒

1927 年
卡地亚巴黎
玫瑰金，黄金，铂金
玫瑰式切割钻石，螺钿，青金石，孔雀石
红色、黑色珐琅
6.40（直径）× 1.20 厘米

由螺钿、金和石制成的中心饰板出自弗拉迪米尔　马科斯基之手（参见VC 37 A27, VC 58 A25）。

277

PB 21 A27

Powder compact

Cartier Paris, 1927
Pink gold, yellow gold, platinum
Rose-cut diamonds, mother-of-pearl, lapis
lazuli, malachite
Red and black enamel
6.40 (diameter) × 1.20 cm

The mother-of-pearl, gold and hard-stone central plaque was created by Vladimir Makowsky.

(see VC 37 A27, VC 58 A25)

278

DI 12 A27

佛犬墨水瓶

1927 年
卡地亚巴黎
金
玫瑰色石英
玛瑙花瓶（中国 19 世纪）
青金石
凸圆形蓝宝石
蓝色珐琅
10.55 × 9.05 × 8.60 厘米

278

DI 12 A27

Fó dog inkwell

Cartier Paris, 1927
Gold
Rose quartz
Carved grey agate vase (Chinese 19th
century)
Lapis lazuli
Sapphire cabochons
Blue enamel
10.55 × 9.05 × 8.60 cm

墨水瓶

约 1927 年
卡地亚纽约
金
红釉瓷
木质雕花
9.15 × 6.30 厘米

售予哈里森·威廉姆斯夫人，即莫娜·俾斯麦伯爵夫人。

美貌动人的莫娜·斯特拉德（1897～1983年）生于肯塔基州路易斯维尔一个非常普通的家庭。在经过两次富有但却并不幸福的婚姻之后，她于1926年嫁给美国巨富哈里森·威廉姆斯。后者在新兴电力行业和一系列成功的股票操纵中赢得了财富。1933年，著名法国时装设计师莫利诺、浪万、维奥内、勒隆和香奈儿共同推选她为"全球衣着最佳女士"。斯特拉德夫人热衷于收集工艺品和珠宝。在哈里森·威廉姆斯先生亡故之后，她嫁给德国首相俾斯麦的孙子爱德华·俾斯麦。

Inkwell

Cartier New York, c. 1927
Gold
Red enamelled porcelain
Carved wood
9.15 × 6.30 cm

Sold to Mrs. Harrison Williams, Countess Mona Bismark

The beautiful Mona Strader (1897-1983), was born in Louisville Kentucky from a very modest family. After two wealthy but unhappy weddings, she married in 1926 the immensely rich American Harrison Williams, who had built his fortune in the electrical power boom and with a series of successful stock manipulations. In 1933, the famous French couturiers Molyneux, Lanvin, Vionnet, Lelong and Chanel got together and voted her the best dressed woman in the world. During her long life, she became a great collector of objets d'art and jewels. After the death of Harrison Williams, she married Comte Edouard Bismarck, grandson of the German Grand Chancellor.

280

DI 08 C27

佛犬镇纸

约 1927 年
卡地亚纽约
金
雕刻象牙（佛犬）
玉（柱身，底座）
凸圆形蓝宝石，宝石
红色珐琅
6.75 × 5.95 × 5.95 厘米

售予哈里森·威廉姆斯夫人，即蒙
娜·俾斯麦伯爵夫人（参见DI 07 C27，
FK 19 A26）。

象牙雕刻的佛犬可追溯至19世纪
下半叶。

280

DI 08 C27

Fô dog Paper-weight

Cartier New York, c. 1927
Gold
Engraved ivory (Fô dog)
Nephrite (column, base)
Sapphire cabochons, precious stones
Red enamel
6.75 × 5.95 × 5.95 cm

Engraved *Fô dog* in ivory from the
second half of the XIX Century.

Sold to Mrs. Harrison Williams,
Countess Mona Bismarck
(see DI 07 C27, FK 19 A26)

EG 11 A28

耳坠（一对）

1928 年

卡地亚纽约

铂金，金

圆形旧式切割和单面切割钻石

翡翠佛像

黑色珐琅

5.9 × 1.85 厘米

281

EG 11 A28

Pair of ear-pendants

Cartier New York, 1928

Platinum, gold

Round old- and single-cut diamonds

Carved jade Buddhas

Black enamel

5.9 × 1.85 cm

282

BT 109 A28

喀迈拉手镯

1928 年

卡地亚巴黎

金，铂金

枕形、单面切割、梨形和长阶梯形切割钻石

两颗刻槽式祖母绿珠、一颗雕花祖母绿、面弧形

祖母绿和凸圆形祖母绿

两颗叶形雕花蓝宝石、凸圆形蓝宝石和面弧形

蓝宝石

雕花珊瑚

绿色、蓝色、黑色珐琅

7.40 × 8.15 × 1.60 厘米

售予·加娜·瓦斯卡（参见CM 09
A23, JA 26 A30）。

倒转枕形钻石代表眼睛。旋转其
中一颗龙头便可打开手镯。此枚手镯
完美结合了印度和中国风格，采用多
种材质制成喀迈拉，是卡地亚的经典
之作。

282

BT 109 A28

Chimera bangle

Cartier Paris, 1928

Gold, platinum

Cushion-shaped, single-, pear- and

baguette-cut diamonds

Two fluted emerald beads, one carved

emerald, buff-top emeralds and emerald

cabochons

Two leaf-shaped carved sapphires, sapphire

cabochons and buff-top sapphires

Carved coral

Green, blue and black enamel

7.40 × 8.15 × 1.60 cm

The eyes are represented by inverted
cushion-shaped diamonds. The
bracelet opens by twisting one of
the two dragon heads. This bangle is
a blend of Indian and Chinese tradi-
tions—produced in various materials,
the *Chimera* bracelet became a great
Cartier classic.

Sold to Ganna Walska
(see CM 09 A23, JA 26 A30)

283

CL 48 A28

中国花瓶胸针

1928 年
卡地亚巴黎
金，铂金
玫瑰式切割钻石
叶形雕花祖母绿
青金石
雕花珊瑚
黑色珐琅
4.82 × 2.90 × 1.25 厘米

284

CL 101 A28

鹦鹉胸针

1928 年
卡地亚巴黎
铂金
单面切割钻石
蓝宝石，面弧形切割和圆柱形切割祖母绿
镶嵌玉石
雕花珊瑚和凸圆形珊瑚
黑色珐琅（眼睛）
4.00 × 1.70 厘米

283

CL 48 A28

Chinese vase brooch

Cartier Paris, 1928
Gold, platinum
Rose-cut diamonds
Leaf-shaped engraved emeralds
Lapis lazuli
Carved coral
Black enamel
4.82 × 2.90 × 1.25 cm

284

CL 101 A28

Parrot brooch

Cartier Paris, 1928
Platinum
Single-cut diamonds
Sapphires, buff-top and calibré-cut emeralds
Engraved jade
Carved coral and coral cabochons
Black enamel (eye)
4.00 × 1.70 cm

化妆盒

1928 年
卡地亚巴黎
铂金，金
玫瑰式切割钻石，凸圆形蓝宝石，祖母绿
螺钿，玉，珊瑚，东菱玉
绿松石，黑玛瑙
黑色珐琅
1.6 × 5.8 × 10.0 厘米

内设一面镜子，两个翻盖隔层和一个唇膏架。

画中的人物坐在石凳上，面前有一张供桌，地上散着一些石头，也许是代表一些摆放在花园中的稀有的供石。画面左侧有竹子枝叶，画中的树看似是松，而花瓶上的图案也像是梅花图。纵观整个画面，取材于中国的"三友"（松、竹、梅）图（参见目录朱迪·罗德《卡地亚1900～1939年》200页）。

这画面源于路易·卡地亚拥有的一件中国瓷器。这个康熙时期的五彩瓷盘从路易·卡地亚之子克洛德·卡地亚的遗产中售出——参见1979年11月25至27日摩纳哥苏富比拍卖行记录，101号（感谢艾伦·卡地亚提供的资料）。

285

VC 72 A28

Vanity case

Cartier Paris, 1928
Platinum, gold
Rose-cut diamonds, sapphire cabochons,
emeralds
Mother-of-pearl, jade, coral, aventurine
quartz
Turquoise, onyx
Black enamel
1.6 × 5.8 × 10.0 cm

The interior fitted with a mirror, two lidded compartments and a lipstick holder.

The source of the scene is a Chinese porcelain plate owned by Louis Cartier. This *famille verte* plate of the Kangxi period was sold from the estate of Louis' son, Claude Cartier, at Sotheby Parke Bernet, Monaco, 25-27 November 1979, lot 101 (reference kindly provided by Alain Cartier).

The figure is seated on a stool in front of a holy table with rocks in the foreground, possibly representing rare rocks often collected to put in gardens as holy rocks. There are bamboo branches on the left, while the tree behind may be a pine and the flowers in the vase may be prunus blossoms. Taken together, these elements are the Three Friends of Chinese mythology. (see catalogue Judy Rudoe - Cartier 1900-1939, p. 200)

286

VC 71 A28

中国花瓶装饰化妆盒

1928 年
卡地亚巴黎
金，铂金
雕花祖母绿，红宝石和蓝宝石
玫瑰式切割钻石
黑玛瑙
珊瑚
奶油色、蓝色、绿色珐琅
1.78 × 5.67 × 8.68 厘米

内设一面镜子，两个翻盖隔层和
一个唇膏架。

286

VC 71 A28

Vanity case with Chinese vase decoration

Cartier Paris, 1928
Gold, platinum
Carved emeralds, rubies and sapphires
Rose-cut diamonds
Onyx
Coral
Cream-coloured, blue and green enamel
1.78 × 5.67 × 8.68 cm

The interior fitted with a mirror, two lidded compartments and a lipstick holder.

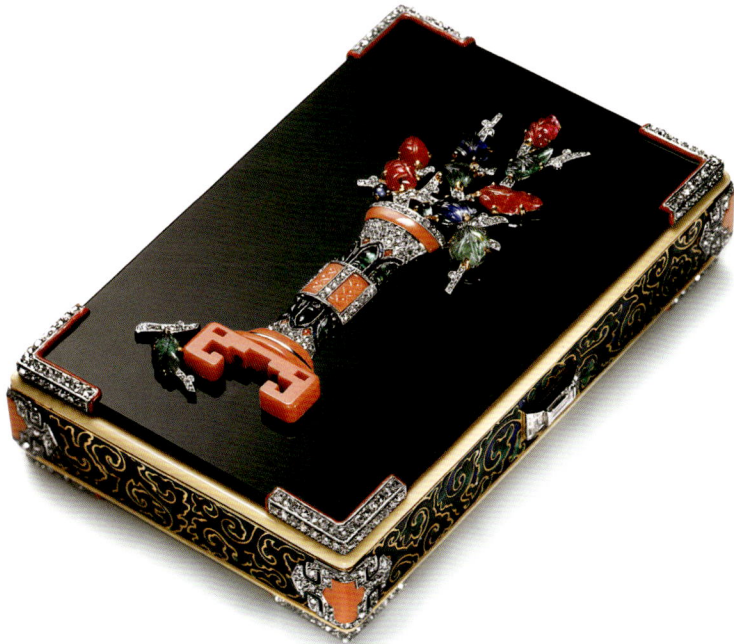

287

PB 20 A28

粉盒

1928 年
卡地亚巴黎
金
雕花翡翠，青金石
凸圆形蓝色宝石
绿色、蓝色珐琅
3.0 × 5.50 × 4.40 厘米

来源于伊拉·尼尔森·莫里斯夫人
（参见VC 65 A27）。

287

PB 20 A28

Powder box

Cartier Paris, 1928
Gold
Carved jade, lapis lazuli
A sapphire cabochon
Green and blue enamel
3.0 × 5.50 × 4.40 cm

Provenance : Mr. Ira Nelson Morris
(see VC 65 A27)

288
DI 28 A28
龙形墨水瓶
1928 年
卡地亚巴黎
金
凸圆形祖母绿
烟色雕花石英（龙）
雕花玛瑙（支架）
雕花珊瑚
12.73 × 11.98 × 10.65 厘米

基座的两侧都可作为抽屉，这些
抽屉设计的原意是放置邮票。

288
DI 28 A28
Dragon inkwell
Cartier Paris, 1928
Gold
Emerald cabochons
Carved smoky quartz (dragon)
Carved agate (stand)
Carved coral
12.73 × 11.98 × 10.65 cm

Each side of the stand composed of a
drawer. These drawers were originally
for stamps.

289

OV 06 A28

佛像陈列装饰品

1928 年
卡地亚巴黎
金，铂金，银
玫瑰色雕花石石英，玉，青金石
蓝宝石、凸圆形红宝石
珍珠
玫瑰式切割钻石
16.5 × 11.7 × 10.0 厘米

佛像的头、舌头以及双手都是活动连接的。

289

OV 06 A28

Buddha cabinet ornament

Cartier Paris, 1928
Gold, platinum, silver
Carved rose quartz, nephrite, lapis lazuli
Sapphire and ruby cabochons
Pearls
Rose-cut diamonds
16.5 × 11.7 × 10.0 cm

The head, the tongue and the hands
are articulated.

290

BT 64 A29

喀迈拉手镯

1929 年
卡地亚巴黎, 特别订制
铂金
梨形钻石, 圆形旧式切割、单面切割及法式切割
钻石
凸圆形蓝宝石, 面弧形蓝宝石, 祖母绿
圆模雕刻装饰无色水晶（喉部）
7.8 × 7.7 × 1.8 厘米

这款手镯是卡地亚喀迈拉系列第
一款全铂金宝石手镯。

290

BT 64 A29

Chimera bracelet

Cartier Paris, special order, 1929
Platinum
Pear-shaped diamonds, round old-, single-
and French-cut diamonds
Sapphire cabochons, buff-top sapphires
and emeralds
Gadrooned rock crystal (throat)
7.8 × 7.7 × 1.8 cm

This bracelet was the first of the
Cartier's *Chimera* bangles to be made
entirely of platinum and gemstones.

291

CDS 06 A29

支架式座钟

1929 年
卡地亚巴黎
镀金金属，铂金
硬橡胶（外壳）
描绘中国风景的螺钿漆画（表盘）
凸圆形月长石，螺钿
珊瑚条，黑色珐琅，玫瑰式切割钻石
8.30 × 8.30 × 2.20 厘米

　　此钟为8日储存圆形机芯，模拟日内瓦波纹装饰，镀铑，3个调校项目，15颗宝石轴承，瑞士杠杆式擒纵结构，双金属平衡摆轮，宝玑摆轮游丝，3点钟位置设可伸缩上链表冠。座钟外壳最初为象牙，1930年出售前改为硬橡胶。

　　卡地亚所使用的大部分漆画都采用的是远东的小型嵌画形式。这些嵌片均采用的是一种被称为螺钿的技法，以黑色漆面内嵌螺钿，创作于19世纪的琉球群岛（中国和日本之间），或是日本本土的螺钿漆画中心长崎。螺钿着粉色和绿色。在将黑色漆面与螺钿填平之后，再最后上一层漆面，将螺钿隐藏起来。然后，再用浮石对表面进行打磨，再度将螺钿显露出来，由此获得极致光滑的表面效果（参见朱迪·罗德目录，《卡地亚1900～1939年》，第336页）。

291

CDS 06 A29

Desk clock with strut

Cartier Paris, 1929
Gilded metal, platinum
Ebonite (case)
Plaque of *laque burgauté* (dial) depicting a
Chinese landscape
Moonstone cabochon, mother-of-pearl
Coral rods, black enamel, rose-cut
diamonds
8.30 × 8.30 × 2.20 cm

Round 8-day movement, fausses Côtes de Genève decoration, rhodium-plated, 3 adjustments, 15 jewels, Swiss lever escapement, bimetallic balance, Breguet balance spring. Retractable winding crown at 3 o'clock.

The case, originally of ivory, was redone in ebonite in 1930, prior to sale.

Much of the lacquer used by Cartier was in the form of small plaques imported from the Far East. These were of black lacquer inlaid with mother-of-pearl, a method known as laque burgauté, and were made in the nineteenth century either on the Ryukyu islands between China and Japan, or in Nagasaki, the main centre of production in Japan itself. The mother-of-pearl was partly tinted in purple and green. After the lacquer had been brought level with the mother-of-pearl inlay, a final lay of lacquer was applied to the whole object, thus concealing the mother-of-pearl. The surface was then rubbed with pumice stone until the inlay was once more exposed. In this way a perfectly smooth surface was obtained.

(see catalogue Judy Rudoe, Cartier 1900-1939, p. 336).

印章表胸针

1929 年
卡地亚巴黎
金，铂金
圆形旧式切割、长阶梯形切割、单面切割和玫瑰
式切割钻石
两颗面弧形凸圆形红宝石，红宝石珠
雕花翡翠
黑玛瑙
黑色、白色珐琅
11.2 × 2.8 × 1.6 厘米

这款表镶嵌了一个19世纪中国翡翠印，印上雕刻佛教中的狮子形象。

此表为积家103型椭圆形机芯，日内瓦波纹装饰，镀铑，2个调校项目，19颗宝石轴承，瑞士杠杆式擒纵结构，双金属平衡摆轮，扁平摆轮游丝。

Seal watch-brooch

Cartier Paris, 1929
Gold, platinum
Round old-, baguette-, single-, and rose-cut
diamonds
Two buff-top ruby cabochons, ruby beads
Carved jade
Onyx
Black and white enamel
11.2 × 2.8 × 1.6 cm

The watch set in a 19th-century Chinese jade seal featuring a Buddhist lion figure.

Oval LeCoultre caliber 103 movement, Côtes de Genève decoration, rhodium-plated, 2 adjustments, 19 jewels, bimetallic balance, Swiss lever escapement, flat balance spring.

鹦鹉胸针

1929 年
卡地亚巴黎
铂金，金
玫瑰式和单面切割钻石
凸圆形蓝宝石和面弧形蓝宝石
一颗凸橄榄形翡翠
珊瑚
黑色珐琅
3.7 × 1.1 × 1.0 厘米

Parrot brooch

Cartier Paris, 1929
Platinum, gold
Rose- and single-cut diamonds
Sapphire cabochons and buff-top sapphires
One jade cabochon
Coral
Black enamel
3.7 × 1.1 × 1.0 cm

294

JA 26 A30

腰带

1930 年
卡地亚伦敦
金
21 颗镶刻红宝石底座
21 块雕花翡翠
长度　85.60 厘米

1930年，卡地亚伦敦将这条腰带售予加娜·瓦斯卡。

翡翠来自中国，雕琢于19世纪末期，模拟清光绪钱币。此类翡翠腰带具有典型的中国特征。在当时的高级官员常将此类腰带佩戴于丝绸长袍之外。红宝石由卡地亚伦敦镶嵌（参见 CM 09 A23、BT 109 A28）。

294

JA 26 A30

Belt

Cartier London, 1930
Gold
Twenty-one engraved collet-set rubies
Twenty-one carved jade disks
Length 85.60 cm

The jade medallions come from China. They were carved in the late nineteenth century to imitate coins from the reign of Emperor Guangxu (Qing Dynasty).

Typically Chinese, this type of jade belt was worn over the long silk robes of high-ranking dignitaries during the Ming dynasty. The rubies were set by Cartier London, where the belt was sold in 1930 to Ganna Walska.

(see CM 09 A23, BT 109 A28)

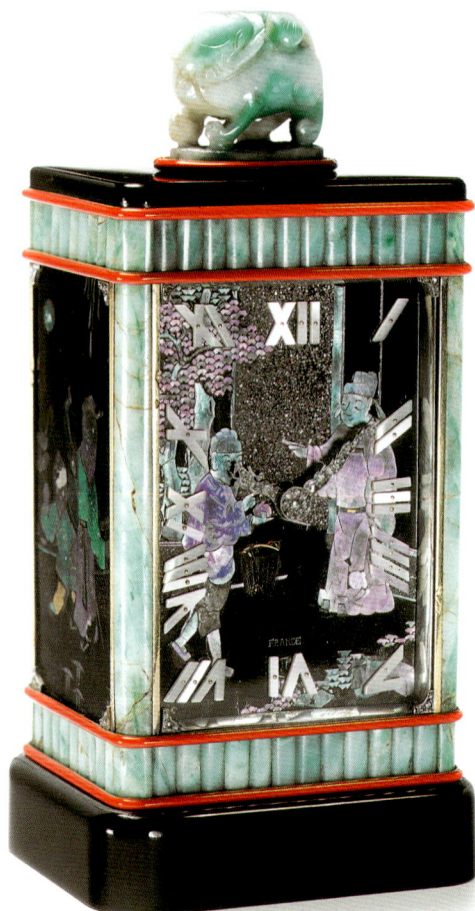

三问座钟

1930 年
卡地亚巴黎
金，铂金
玫瑰式切割钻石
两幅描绘中国风景的螺钿漆画
雕花翡翠，螺钿，黑玛瑙
红色、黑色珐琅
高度　12.0 厘米

此钟为8日储存长方形机芯，三问功能，镀金，3个调校项目，11颗宝石轴承，瑞士杠杆式擒纵结构，双金属平衡摆轮，宝玑摆轮游丝。底座带机芯上链和时间调校钥匙（参见CDS 06 A29）。

295

CR 17 A30

Desk clock with minute repeater

Cartier Paris, 1930
Gold, platinum
Rose-cut diamonds
Two plaques of laque burgauté depicting
Chinese scenes
Carved jade, mother-of-pearl, onyx
Red and black enamel
Height 12.0 cm

Rectangular 8-day movement, minute repeater, gold-plated, 3 adjustments, 11 jewels, bimetallic balance, Swiss lever escapement, Breguet balance spring. Key for winding movement and setting hands, housed under the base.

(see CDS 06 A29)

296

CDB 10 A30

佛犬座钟

1930 年
卡地亚巴黎
金, 铂金, 银镀金
玫瑰式切割钻石
凸圆形红宝石
两个翡翠佛犬
螺钿
黑玛瑙, 无色水晶, 珊瑚
黑色, 红色珐琅
9.5 × 8.1 × 5.32 厘米

此钟为8日储存圆形机芯, 模拟日内瓦波纹装饰, 镀铑, 3个调校项目, 15颗宝石轴承, 瑞士杠杆式擒纵结构, 双金属平衡摆轮, 宝玑摆轮游丝。

296

CDB 10 A30

Desk clock with *Fô dogs*

Cartier Paris, 1930
Gold, platinum, silver-gilt
Rose-cut diamonds
Ruby cabochons
Two carved jade *Fô Dogs*
Mother-of-pearl
Onyx, rock crystal, coral
Black and red enamel
9.5 × 8.1 × 5.32 cm

Round 8-day movement, fausses Côtes de Genève decoration, rhodium-plated, 3 adjustments, 15 jewels, Swiss lever escapement, bimetallic balance, Breguet balance spring.

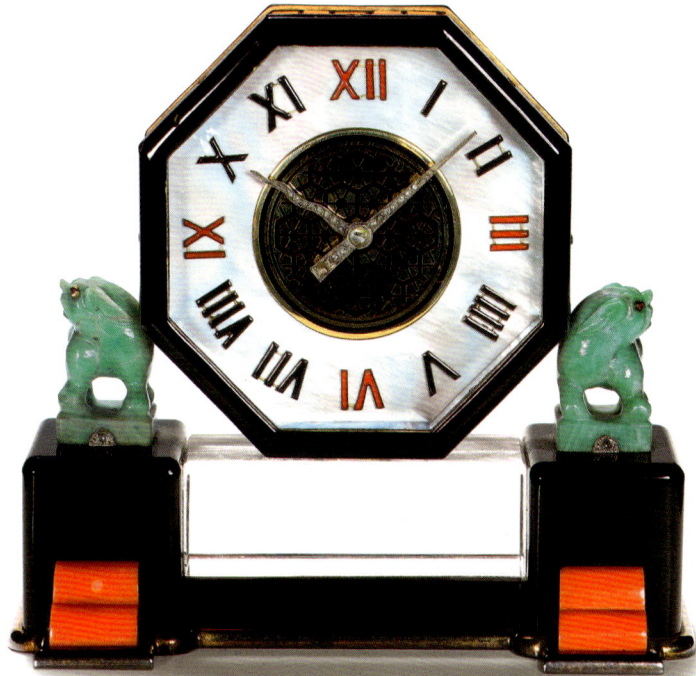

297

CM 04 A31

神像报时魅幻钟

1931 年
卡地亚巴黎
铂金，金
雕花白玉
无色水晶（钟盘），黑玛瑙（支架），玉（底座）
玫瑰式切割钻石，珍珠，凸圆形绿松石，珊瑚
深蓝色、绿蓝色、红色、黑色珐琅
35.0 × 28.0 × 14.0 厘米

售予保罗·路易·威勒。

此钟为8日储存长方形机芯，报时机制（小时和刻钟），镀金，瑞士杠杆式擒纵结构，双金属平衡摆轮，宝玑摆轮游丝。上弦轴柄和时间调校轴柄设于底座后。

此款魅幻时钟是以动物和小雕像为主题的时钟系列12款中的最后一款（参见CS11A25, CM23A26）。

保罗·路易·威勒(1893～1993年)是19世纪一个法国大家族的后裔。他是一名杰出的工程师、第一次世界大战中的王牌飞行员、航空工业的先驱、国际金融家，并且还是一个艺术赞助者。作为卡地亚的大客户，威勒于1937年购买了一颗重达245.35克拉的纪念版钻石。这颗钻石是当时全球第四大的钻石。

297

CM 04 A31

Striking mystery clock with *deity*

Cartier Paris, 1931
Platinum, gold
Carved white jade
Rock crystal (dial), onyx (stand), nephrite (base)
Rose-cut diamonds, pearls, turquoise cabochons, coral
Dark-blue, turquoise-blue, red and black enamel
35.0 × 28.0 × 14.0 cm

Rectangular 8-day movement, striking mechanism (hours and quarter-hours), gold-plated, Swiss lever escapement, bimetallic balance, Breguet balance spring. Arbor for winding movement and setting hands on back of base.

This mystery clock was the last in a series of 12 that featured animals or figurines.

Sold to Paul Louis Weiller

Paul Louis Weiller (1893–1993) was the scion of a grand 19th century French family. An engineer, ace pilot during World War I, pioneer of the aeronautic industry, and international financier, he was also a great art patron. A major Cartier client, in 1937 Weiller notably bought the 245.35-carat Jubilee diamond, the world's fourth largest at the time.

(see CS 11 A25, CM 23 A26)

413

298

IO 56 A31
桌铃

1931 年
卡地亚伦敦
金
玫瑰色石英, 玉
9.0 × 5.4 × 5.0 厘米

298

IO 56 A31

Table bell

Cartier London, 1931
Gold
Rose quartz, nephrite
9.0 × 5.4 × 5.0 cm

299

RG 30 A34
戒指

1934 年
卡地亚巴黎, 特别订制
金
面弧形和圆柱形切割红宝石, 长阶梯形切割钻石
一颗凸圆形翡翠 (重 37.67 克拉, 翡翠来自客人所提供的另一枚戒指)
2.90 × 2.07 厘米

售予芭芭拉·赫顿。
芭芭拉·赫顿 (1912～1979年) 是伍尔沃斯零售连锁店创始人之孙女, 也是全球最富有的女性之一。1933年6月23日, 她嫁给俄罗斯王子阿历克斯·米德瓦尼——她的7位丈夫中的第一位。在婚礼上, 她佩戴了一条特殊的珍珠项链和镶钻玳瑁冠冕, 均为卡地亚制作。她钟爱富丽堂皇的珠宝, 是卡地亚最忠实的客人之一 (参见CL 140 A57)。

299

RG 30 A34

Ring

Cartier Paris, special order, 1934
Gold
Buff-top and calibré-cut rubies, baguette-cut diamonds
One 37.67 carat jade cabochon
The jade comes from another ring supplied by the client
2.90 × 2.07 cm

Sold to Barbara Hutton
Barbara Hutton (1912–1979) was the granddaughter of the founder of the Woolworth's retail chain, and one of the richest women in the world. On June 23, 1933, she married the Russian prince Alexis Mdivani, the first of her seven husbands. For her wedding, she wore an extraordinary pearl necklace and a diamond-studded tortoiseshell tiara, both made by Cartier. An enthusiast for magnificent jewellery, she became one of the firm's most loyal clients.

(see CL 140 A57)

300

SA 04 A35

烟灰缸

1935 年
卡地亚巴黎
银
玉（睡莲）
黑曜石（底座）
雕花珊瑚
玫瑰式切割钻石
黑色漆面
5.60 × 9.0 × 5.60 厘米

300

SA 04 A35

Ashtray

Cartier Paris, 1935
Silver
Carved jade (water lily)
Obsidian (base)
Carved coral
Rose-cut diamonds
Black lacquer
5.60 × 9.0 × 5.60 cm

301

LR 22 A39

台式打火机

1939 年
卡地亚巴黎
银，金，铂金
玉
珊瑚
圆形旧式切割钻石
黑色珐琅
6.5 × 4.5 × 2.0 厘米

301

LR 22 A39

Table lighter

Cartier Paris, 1939
Silver, gold, platinum
Engraved jade
Coral
Round old-cut diamonds
Black enamel
6.5 × 4.5 × 2.0 cm

302

CS 05 A43

瑞兽时钟

1943 年
卡地亚巴黎
金,铂金,银
雕花珊瑚
白色玛瑙(底座和钟盘)
祖母绿,旧式切割钻石
黑色珐琅
21.0 × 14.0 × 8.0 厘米

此钟为8日储存雷玛尼亚圆形机芯,垂直式刻槽,镀铑,3个调校项目,15颗宝石轴承,瑞士杠杆式擒纵结构,双金属平衡摆轮,宝玑摆轮游丝。调校钟冠设于底座后部。时针由外圈中的旋转盘驱动,指向珊瑚宝石杆,以示小时。

此款时钟制于第二次世界大战物资匮乏的时期。它重新使用了1928年拆解时钟上的19世纪中国珊瑚小雕像。

302

CS 05 A43

Turtle-Chimera clock

Cartier Paris, 1943
Gold, platinum, silver
Carved coral
White onyx (base and dial)
Emeralds, old-cut diamonds
Black enamel
21.0 × 14.0 × 8.0 cm

Round Lemania 8-day movement, vertical reeding, rhodium-plated, 3 adjustments, 15 jewels, Swiss lever escapement, bimetallic balance, Breguet balance spring. Winding crown on back of base. The hour pointer is driven by a disk that turns within the hoop, pointing to the coral rods as hour markers.

Made during the shortages of World War II, this clock re-used a 19th-century Chinese coral figurine taken from a dismantled clock of 1928.

303

DI 27 A52

打火机 / 镇纸

1952 年
卡地亚巴黎
银，金
雕花玉（佛手图案）
珊瑚，玫瑰式切割钻石
黑色漆面
9.39 × 3.62 × 4.37 厘米

303

DI 27 A52

Lighter/Paper-weight

Cartier Paris, 1952
Silver, gold
Carved gray jade (Buddha's hand motif)
Coral, rose-cut diamonds
Black lacquer
9.39 × 3.62 × 4.37 cm

自 19 世纪开始，卡地亚销售记录中就出现了"埃及"和"印度"珠宝。然而，当时的卡地亚仍与其他珠宝商货源相同，所售卖的珠宝也基本大同小异。直至 1910 年，珠宝和配饰才在古代东方文明的影响之下，呈现出一种全然不同的风格。一系列新的考古发现，引发了公众对埃及和美索不达米亚文化的痴迷，也加剧了这一风格的演变。

折衷主义的装饰艺术风格，其灵感主要来自于古老的异域文化，尤其是非洲与东方文明。在这些艺术灵感的激发下诞生的全新造型与图案设计，大多被用于建筑、室内设计、家具、织物之中，珠宝自然也不例外。在埃及、波斯、印度和远东艺术的影响下，卡地亚设计师设计出一系列不拘一格、精妙绝伦的作品。这些作品多为珠宝和工艺品，大致可分为两类：一类以贵金属和宝石制成；一类以古董、珠宝重新组合而成。这种将古代和近代的非欧洲艺术与当代风格相结合的创造性构思，为卡地亚带来了令人叹为观止的创作成果，并在 20 世纪 20 年代的装饰艺术思潮中独树一帜，成为卡地亚的特色之一。

埃及。20 世纪初，由于考古发现，埃及艺术成为了美术、时尚、文学和电影的灵感之源。路易·卡地亚在研究了地中海艺术之后，产生了将古代珠宝和工艺品残片用当代手法重新镶嵌的构想，并对这一构想非常执著，时常去拜访巴黎两大专营古代地中海珍玩的艺术品经销商——卡布耶兄弟和迪克兰·科勒凯，甚至还亲自向卢浮宫专家寻求建议。卡地亚极为看重古董雕刻或雕塑元素，并以巧妙的手法对其加以强调。设计师以黑玛瑙、宝石和珊瑚来勾勒轮廓，比如在简单的圣甲虫彩陶片四周，装上奢华的宝石翅膀。众多优美作品也由此诞生。如精心复制出象形文字的埃及自鸣钟（参见 CDB 21 A27），圣甲虫胸针（参见 CL 264 A25）、埃及神胸针（参见 CL 278 A25, CL 161 A27）等，均为只此一件的孤品。1997 年，这批作品在大英博物馆展出时专家们对其中的组成碎片进行鉴定，确定它们来自于公元前 2000 年（参见 CL 263 A25, VC 65 A27）。

波斯。1913 年，卡地亚在纽约举行名为"卡地亚珠宝创作——来自印度、波斯、阿拉伯、俄国和中国的艺术"展。展出的 50 件作品，半数以上为波斯或印度风格，在展品目录的封面还绘制了波斯书籍的装帧方式。这是一次具有历史纪念意义的展览，充分印证了波斯和印度文明对卡地亚作品的影响。路易·卡地亚是波斯细密画和古代手稿的著名收藏家，由于伊斯兰宗教古迹中没有人形和动物形的象征图案，艺术家们在数百年间发展了无数的线形装饰手法，主要是各种风格的花叶图案，包括精美的藤蔓。这些藏品正是精美配饰及钟表装饰的重要灵感来源。在他辞世之后，这些藏品也随之被拍卖，其中一部分至今仍保留在哈佛大学法格艺术馆。

印度。对卡地亚珠宝的异国风格影响最大的国家无疑是印度。受大英帝国的影响，印度在伦敦远甚于其在巴黎的影响。1909 年，执掌伦敦分部的阿尔弗雷德·卡地亚的幼子雅克受命处理所有与印度相关的业务：宝石采购，与印度客户和王公贵族建立联系。1911 年，雅克到印度旅行，他惊奇地发现，印度王公非常痴迷于巴黎人设计的珠宝和腕表。他们要求卡地亚将自己的家传珠宝改造为现代款式，用白色的铂金取代已经过时的黄金基座。同时，欧洲人也对印度人的宝石镶嵌和祖母绿式宝石切割法产生了浓厚兴趣，因此，卡地亚的设计师顺应这一风潮创作出高贵华丽的印度风格项链（参见 NE 25 A32, NE 27 A30），或简单地将古代印度珠宝进行重新镶嵌（参见 NE 34 A28, CL 241 A39, CL 58 A38），并因此风靡一时。而最为出神入化，同时也最受追捧的卡地亚珠宝，当属"水果锦囊"风格系列作品，以雕花蓝宝石、红宝石、绿宝石和钻石（参见 BT 12 A28, CL 31 A29）组成，精美绝伦。1936 年，被誉为"全球最优雅"的女性黛丝·法罗斯委托卡地亚制造一条项链（参见 NE 28 A36），为印度王室风格，中间是两颗大的蓝宝石，几经重镶，仍是卡地亚最具代表性的作品之一。

卡地亚艺术典藏室馆长　帕斯卡尔·勒博

Since the 19th century, one can locate sales of Egyptian and Indian jewellery in the Cartier register. However, until that time, Cartier, and most other jewellers, had the same suppliers, thus jewels fundamentally looked the same from one house to another. It wasn't until 1910 that jewels and accessories were created in an entirely different style, with an influence from the ancient Oriental civilizations. New archaeological discoveries causing public fascination with Egyptian and Mesopotamian cultures would amplify this phenomenon.

The eclectic Art Deco style was largely inspired by these exotic cultures of the past, in particular that of Africa and the Orient, inspiring new shapes and motifs often used in architecture, interior design, furniture, fabrics, and of course in jewellery. Influenced as well by Egyptian, Persian, Indian, and Far Eastern arts, Cartier's designers would dream up the most imaginative and wonderful creations which can be divided into two categories: jewels and "objets d'art" made of precious metals and stones; and those incorporating fragments of antique jewels. This inventive concept of fusing ancient and non-European art of earlier epochs with modern styles led Cartier to an astonishing synthesis that was unique to the Art Deco movement of the 1920s and became one of Cartier's specialties.

Egypt

Egypt once more became an inspiration for fine arts, fashion, literature and cinema with the archaeological discoveries of the early 20th Century. Louis Cartier studied the arts of Mediterranean civilizations and brought forth the idea of using fragments of ancient jewellery and "objets d'art" mounted in contemporary settings. Louis took this very seriously, becoming a regular visitor of two of the most important art dealers specialized in treasures of the ancient Mediterranean world: the Kalebdjian brothers and Dikran Kelekian, both established in Paris' Rue de la Paix and Place Vendôme. He would even go to seek advice from specialists at the Musée du Louvre. Cartier was extremely respectful with engraved or sculpted antique elements, wonderfully accenting them: the designers would articulate the outline with onyx, precious stones and coral; for example around a simple faience scarab, sumptuous precious stone wings would spread forth. This led to gorgeous creations, such as the Egyptian Clock, where hieroglyphs were carefully reproduced (see CDB 21 A27), the Scarab brooch (see CL 264 A25), and brooches with divinities (see CL 278 A25, CL 161 A27), each unique and one of a kind. In 1997, some of these pieces were exhibited at the British Museum, where specialists were able to authenticate and date them as being from the second half of the first millennium BC (see CL 263 A25, VC 65 A27).

Persia

The most monumental event celebrating the Persian and Indian influences was the *Collection of Jewels… Created by Messieurs Cartier… from the Hindoo, Persian, Arab, Russian and Chinese Arts* Exhibition held by Cartier in New York in 1913. Fifty jewels, of which more than half were of Persian or Indian origin, were exhibited, the catalogue illustrating the pieces depicting the bindings of a Persian book. An exhibition of historical significance, this fully testifies the influence of Persian and Indian civilization on Cartier works. Throughout the years, Louis Cartier became a renowned collector of Persian miniatures and ancient manuscripts. Anthropomorphous and animal representations being excluded from religious Islamic monuments, the artists throughout the centuries developed an infinite number of linear decorations and stylized vegetal motifs, including the delicate arabesques, inspiring adornments on precious accessories and clocks. After he passed away, Louis Cartier's collection was auctioned off, with some pieces still kept today at the Fogg Art Museum of Harvard University. It is a certainty that this collection was a great source of inspiration to Louis and his designers.

India

The strongest influence on Cartier's exotic jewellery is, without a doubt, India.

Because of the British Empire the influence of India was stronger in London than in Paris, and as soon as 1909, Jacques, the youngest brother who directed the London branch, was entrusted with all the parts of the business with India: stone purchases, contacts with Indian customers, and agreements with the Maharajahs. In 1911, he travelled to India and was very surprised to discover the Maharajahs lust for Parisian designed jewels and watches. They requested Cartier to transform their family treasures into modern creations, replacing the old-fashioned yellow gold mountings with white platinum. At the same time, a passion for gem-set and enameled jewels of Indian inspiration began in Europe, leading Cartier's designers to create magnificent Indian-style necklaces (see NE 25 A32, NE 27 A30), or simply remounted fragments of antique Indian jewels (see NE 34 A28, CL 241 A39, CL 58 A38). However, the definitive and most sought after Cartier jewels, which came to be known as the Tutti Frutti style, were extraordinary pieces composed of carved sapphires, rubies, emeralds, and diamonds (see BT 12 A28, CL 31 A29). A breathtaking piece of this period is a necklace dating from 1936, commissioned by one of the world's most elegant women: the Honorable Daisy Fellowes (See NE 28 A36). Originally designed in the style of a maharajah's necklace, with a cord fastening, and with the two large carved sapphires set at the centre, it underwent several resettings before attaining its current form. This necklace remains one of the most iconic creations of Cartier.

Pascale Lepeu
Curator of the Cartier Collection

荷露斯胸针

1925 年
卡地亚巴黎
铂金，金
圆形旧式切割和单面切割钻石
一颗凸圆形祖母绿
长条形、凸圆形珊瑚
黑玛瑙钮钉和凸圆形黑玛瑙
蓝色埃及彩陶
黑色、红色珐琅
4.50 × 7.1 厘米

　　来自公元前2000年的荷露斯头，最初可能是用于陪葬的"黄金猎鹰衣领"的尾饰。荷露斯通常被绘制为一只猎鹰，或是一个猎鹰的头。荷露斯在悠久的埃及历史中扮演着众多角色：天庭之神，光之神（他的眼睛分别代表太阳和月亮），以及法老的守护神。法老本人也有"荷露斯"之称，被尊为荷露斯在俗世的化身。

304

CL 263 A25

Horus brooch

Cartier Paris, 1925
Platinum, gold
Round old- and single-cut diamonds
One emerald cabochon
Coral rods and cabochons
Onyx studs and cabochons
Blue Egyptian faience
Black and red enamel
4.50 × 7.1 cm

The *Horus* head dates from the second millennium B.C.E. It would originally have been a terminal on a "falcon collar of gold" that was traditionally taken into the other world. *Horus* was usually depicted as a falcon, or the head of a falcon. He played several roles throughout Egypt's long history: god of the heavens, god of light (his eyes represented the sun and moon), and patron deity of pharaohs. The pharaoh—himself called *"Horus"*—was worshipped as a manifest sign of the god's presence on earth.

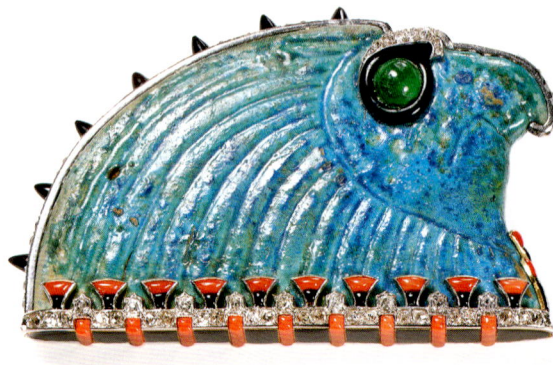

305

CL 264 A25

圣甲虫胸针

1925 年
卡地亚伦敦
金，铂金
圆形旧式切割和单面切割钻石
凸圆形红宝石、祖母绿、黄水晶、黑玛瑙
蓝色古埃及彩陶
12.43 × 5.51 × 1.90 厘米

依照最初的设计，这枚胸针也可
作为腰带扣。

305

CL 264 A25

Scarab brooch

Cartier London, 1925
Gold, platinum
Round old- and single-cut diamonds
Ruby, emerald, citrine and onyx cabochons
Blue ancient Egyptian faience
12.43 × 5.51 × 1.90 cm

Originally, this brooch could also be
worn as a belt buckle.

306

CL 278 A25

狮头女神胸针

1925 年
卡地亚巴黎, 特别订制
金, 铂金
凸圆形红宝石, 祖母绿, 圆形旧式切割和单面切
割钻石
三颗凸圆形黑玛瑙
蓝色埃及彩陶
黑色珐琅
8.30 × 3.67 厘米

　　塞赫迈特, 令人惧畏的狮头女神, 化身自太阳的灼热之眼, 只有经过特别的祭拜才能得到安抚。她会在战争中庇佑法老, 伴随其左右。
　　这尊公元前500～1000年的小雕像, 由卡地亚自巴黎古董商、有地中海艺术专家美誉的卡布耶处购得。

306

CL 278 A25

Sekhmet brooch

Cartier Paris, special order, 1925
Gold, platinum
Ruby and emerald cabochons, round old-
and single-cut diamonds
Three onyx cabochons
Blue Egyptian faience
Black enamel
8.30 × 3.67 cm

Sekhmet, the dreaded lion-headed goddess, is an incarnation of the scorching eye of the sun, which must be appeased through a special rite. She protected the pharaoh and accompanied him in war. Cartier acquired this little statuette, dating from the second half of the first millennium B.C.E., from the Paris dealer Kalebdjian, a great specialist in Mediterranean art.

307

VC 70 A25

埃及石棺化妆盒

1925 年
卡地亚巴黎
黄金和玫瑰金，铂金
玫瑰式切割钻石，祖母绿，圆柱形切割
面弧形蓝宝石
骨制雕花盒盖
橘黄色、蓝色、象牙色珐琅
3.2 × 4.0 × 15.0 厘米

售予乔治·布鲁门塔尔夫人（参见
CDB 21 A27）。

内装一面折叠镜、一把梳子和一
个唇膏架。

307

VC 70 A25

Egyptian sarcophagus
Vanity case

Cartier Paris, 1925
Yellow and pink gold, platinum
Rose-cut diamonds, emeralds, calibré-cut
buff-top sapphires
Engraved bone lid
Saffran, blue and ivory color enamel
3.2 × 4.0 × 15.0 cm

The interior fitted with a folding mirror, a comb and a lipstick holder.

Sold to Mrs George Blumenthal
(see CDB 21 A27)

莲花神胸针

1927 年
卡地亚巴黎为卡地亚纽约制作, 特别订制
铂金, 金
圆形旧式切割钻石
凸圆形祖母绿和红宝石
凸圆形黑玛瑙、圆柱形切割黑玛瑙
蓝色埃及彩陶
黑色、半透明绿色珐琅
5.1 × 3.6 × 2.4 厘米

这个小头来自赛伊斯时期 (公元前6世纪), 是伊希斯雕像的一部分。卡地亚将其镶嵌于一束莲花之间, 大概是想要仿制香气之神奈夫图。他象征着从原初之水中升起的莲花。

根据埃及创世神话, 莲花代表着日出时初生的幼年之神的摇篮。夜晚, 花瓣闭合。次日清晨, 待太阳神重生之时, 莲花方才再度开放。

308

CL 161 A27

Lotus flower deity brooch

Cartier Paris for Cartier New York, special order, 1927
Platinum, gold
Round old-cut diamonds
Emerald and ruby cabochons
Onyx cabochons and calibré-cut onyxes
Blue Egyptian faience
Black and translucent green enamel
5.1 × 3.6 × 2.4 cm

This little head, dating from the Sais period (sixth century B.C.E.) is from a statue of Isis. By setting it on a bouquet of stylized lotus flowers, Cartier probably intended to imitate statuettes of the god Nefertum, lord of perfumes. He symbolized the lotus emerging from the primal waters, which, according to one creation myth, served as a scented cradle for the infant-god representing the first sunrise. In the evening the buds close upon the god, only to reopen again in the morning for his rebirth.

309

JA 19 A27

帽式胸针

1927 年
卡地亚巴黎
铂金，金
单面切割钻石
翡翠
黑玛瑙
3.20 × 2.30 厘米

309

JA 19 A27

Hat brooch

Cartier Paris, 1927
Platinum, gold
Single-cut diamonds
Carved jade
Onyx
3.20 × 2.30 cm

310

CDB 21 A27

埃及自鸣钟

1927 年
卡地亚巴黎
金，银镀金
螺钿嵌片，雕刻象形文字
珊瑚雕花圆环和框架
镶嵌祖母绿、红玉髓和珐琅的埃及神祇
青金石（基座和顶部）
彩色、白色珐琅
24.00 × 15.70 × 12.70 厘米

售予布鲁门塔尔夫人。

此钟为8日储存长方形机芯，自鸣装置（整点和每刻），镀金，嵌珠装饰，3个调校项目，15颗宝石轴承，标准擒纵，双金属平衡摆轮，宝玑摆轮游丝。

乔治·布鲁门塔尔（1858～1941年）是华尔街投资银行拉扎德公司掌门人，自1905年起担任大都会博物馆理事。这款自鸣钟由他的第一任妻子佛罗伦斯（1873～1930年）购得。作为收藏家和众多法国艺术家的资助人，布鲁门塔尔先生和夫人拥有卡地亚的数款重要作品。1935年，在出任大都会博物馆董事会主席。次年，布鲁门塔尔先生迎娶了玛丽·安·佩恩。

310

CDB 21 A27

Egyptian Striking Clock

Cartier Paris, 1927
Gold, silver-gilt
Mother-of-pearl plaques carved with hieroglyphs
Carved coral rings and rods
Egyptian Deity set with emerald, cornelian and enamel
Lapis lazuli (base and top)
Polychrome and white enamel
24.00 × 15.70 × 12.70 cm

Rectangular 8-day movement, striking mechanism (hour and quarter-hours), gold-plated, beaded decoration, 3 adjustments, 15 jewels, platform escapement, bimetallic balance, Breguet balance spring.

Sold to Mrs. Blumenthal

George Blumenthal (1858-1941) headed Lazard, a Wall Street investment bank, and was a trustee of the Metropolitan Museum of Art from 1905 onward. This clock was sold to his first wife, Florence (1873-1930). As collectors and patrons of many French artists, the couple owned several major pieces by Cartier. In 1935, one year after he was appointed president of the Metropolitan Museum, Blumenthal married Mary Ann Payne.

CARTIER

FRANCE

433

311

VC 65 A27

化妆盒

1927 年
卡地亚巴黎
金, 铂金
埃及方解石嵌片
凸圆形祖母绿, 玫瑰式切割钻石
珊瑚, 青金石
绿松石色、蓝色珐琅
9.85 × 5.20 × 2.10 厘米

售予尼尔森·莫里斯。

铰链式盒盖, 内嵌化妆镜。1929年改装为雪茄盒(去除内部隔层和唇膏架)。

据大英博物馆的朱迪·鲁德考证, 侧面所使用的是埃及方解石嵌片。正面是孩提时的荷露斯, 以救世主形象出现, 手持蝎子和狮子, 两侧各有一个书写着象形文字的垂直立柱, 背后有一个魔法咒语。两个雕花嵌片是一个埃及碑石或后世(约前500年)的一块护身符的前后两面, 嵌片正面带浮雕图案, 背面镌刻祈求荷露斯庇护的魔法咒语。类似咒语通常用于免遭毒物叮咬。

尼尔森·莫里斯(1875～1945年), 外交官及作家, 生于芝加哥, 1898年迎娶康斯坦斯·莉莉·罗斯柴尔德(参见PB 20 A28)。

311

VC 65 A27

Vanity case

Cartier Paris, 1927
Gold, platinum
Egyptian calcite plaques
Emerald cabochons, rose-cut diamonds
Coral, lapis lazuli
Turquoise-blue enamel
9.85 × 5.20 × 2.10 cm

The interior of the hinged cover fitted with a mirror. Transformed into a cigarette case in 1929 (interior compartments and lipstick holder removed).

According to Judy Rudoe from the British Museum, the sides are applied with incised Egyptian calcite plaques. On the front Horus the Child, as Saviour, carrying a scorpion and a lion, with a vertical column of hieroglyphs down each side, on the back a magical spell. The two carved plaques originally formed back and front of an Egyptian cippus or amulet of the late period, c. 500 B.C.E., with raised relief in front of a plaque the reverse inscribed with a magical incantation invoking the protection of Horus. Such spells were usually designed to protect against poisonous bites.

Sold to Ira Nelson Morris

Diplomat and author (1875 - 1942), born in Chicago, he married Constance Lily Rothschild in 1898.

(see PB 20 A28)

312

NE 12 A31

项链

1931 年
卡地亚伦敦，特别订制
铂金
单面切割和玫瑰式切割钻石
天然珍珠
花式切割玛瑙
蓝色埃及彩陶
项链长度　38.0 厘米
吊坠　2.7 × 2.1 厘米

项链镶嵌了古埃及蓝色彩陶片。

312

NE 12 A31

Necklace

Cartier London, special order, 1931
Platinum
Single- and rose cut diamonds
Natural pearls
Fancy-cut onyx, blue Egyptian faience
Necklace length 38.0 cm; pendant 2.7 × 2.1 cm

The pieces of blue-glazed faience are ancient Egyptian.

313
CL 277 A13

胸针

1913 年
卡地亚巴黎
铂金
圆形旧式切割和单面切割钻石
种子式镶嵌
9.77 × 2.91 厘米

售予康内留斯·范德比尔特先生。

康内留斯·范德比尔特三世
（1873～1942年）是康内留斯·范德比
尔特之曾孙（参见CL 202 A10）。

313
CL 277 A13

Brooch

Cartier Paris, 1913
Platinum
Round old- and single-cut diamonds
Millegrain setting
9.77 × 2.91 cm

Sold to Mr. Cornelius Vanderbilt

Mr. Cornelius Vanderbilt III (1873-
1942) was the great-grandson of
"Commodore" Cornelius Vanderbilt.

(see CL 202 A10)

314

VC 85 A13

化妆盒

1913 年
卡地亚巴黎
玫瑰金、黄金、铂金
玫瑰式切割钻石
珍珠
黑玛瑙
中央嵌波斯瓷片，装饰彩色珐琅
4.5×7.2 厘米

来源于黛丝·法罗斯（参见BT 44
A21，NE 28 A36，EG 28 A63）。

内设一个带翻隔层和两个唇膏
架。

314

VC 85 A13

Vanity case

Cartier Paris, 1913
Pink gold, yellow gold, platinum
Rose-cut diamonds
Pearls
Onyx
Multicolored enamel on central Persian
plaque
4.5 × 7.2 cm

The interior fitted with a lidded compartment and two lipstick holders.

Provenance: Daisy Fellowes

(see BT 44 A21, NE 28 A36, EG 28
A63)

315

CC 80 A13

香烟盒

1913 年
卡地亚巴黎
金，铂金
玫瑰式切割钻石
黑色、白色珐琅
黑玛瑙
9.20 × 4.41 × 1.80 厘米

盒盖装饰雕花马眼形水晶嵌片，黑色绘花。内设香烟盒和火柴盒，盒盖带划火面。

315

CC 80 A13

Cigarette case

Cartier Paris, 1913
Gold, platinum
Rose-cut diamonds
Black and white enamel
Onyx
9.20 × 4.41 × 1.80 cm

The cover decorated with an engraved navette-shaped crystal plaque and painted in black.

The interior fitted with cigarette and match compartments with a striking surface in the cover.

支架式座钟

1920 年
卡地亚巴黎
铂金，金，镀金金属
玫瑰式切割钻石
凸圆形红宝石
硬橡胶
7.95 × 7.95 厘米

售予玛尔伯勒公爵夫人。

此钟为8日储存圆形机芯，镀金，瑞士杠杆式擒纵结构，双金属平衡摆轮，宝玑摆轮游丝。

Desk clock with strut

Cartier Paris, 1920
Platinum, gold, gilded metal
Rose-cut diamonds
Ruby cabochons
Ebonite
7.95 × 7.95 cm

Round 8-day movement, gold-plated, Swiss lever escapement, bimetallic balance, Breguet balance spring.

Sold to the Duchess of Marlborough

3¹⁷

HO 05 A23

束发带

1923 年
卡地亚巴黎，特别订制
铂金
枕形和圆形旧式切割钻石
中央高度 7.5 厘米

这个束发带可部分拆开，组成两
个带式手镯。

3¹⁷

HO 05 A23

Bandeau

Cartier Paris, special order, 1923
Platinum
Cushion-shaped and round old-cut
diamonds
Height at centre 7.5 cm

Parts of this bandeau can be detached to form two strap bracelets.

318

EG 06 A24

耳坠（一对）

1924 年
卡地亚伦敦
铂金
马眼形和圆形旧式切割钻石
6.75 × 1.35 × 0.38 厘米

售予哈考特子爵夫人。

1899年，美国人玛丽·埃塞尔·彭斯，即J·皮尔彭·摩根之侄女，嫁给英国政治家刘易斯·哈考特子爵。后者曾于1910～1915年期间出任殖民地事务大臣。

318

EG 06 A24

Pair of ear-pendants

Cartier London, 1924
Platinum
Marquise- and round old-cut diamonds
6.75 × 1.35 × 0.38 cm

Sold to the Viscountess Harcourt

In 1899, the American Mary Ethel Burns, niece of J. Pierpont Morgan, married Viscount Lewis Harcourt, a British politician who served as secretary of state for the colonies from 1910 to 1915.

胸针

1924 年
卡地亚巴黎
铂金，金
圆形旧式切割、单面切割钻石
一颗凸圆形蓝宝石（重约 57.60 克拉）
磨砂无色水晶
珍珠
螺钿
黑色珐琅
9.3 × 3.7 厘米

这颗凸圆形蓝宝石来自斯里兰卡，未经过热加工。

这枚胸针最初使用的是一颗雕花祖母绿，而不是凸圆形蓝宝石。

Brooch

Cartier Paris, 1924
Platinum, gold
Round old- and single-cut diamonds
One sapphire cabochon, weighing
approximately 57.60 carats
Frosted rock crystal
Pearls
Mother-of-pearl
Black enamel
9.3 × 3.7 cm

The sapphire cabochon is certified as being from Sri Lanka and never having undergone heat treatment.

This brooch was originally made with a carved emerald in place of the sapphire cabochon.

320

VC 34 A24

化妆盒

1924 年
卡地亚巴黎
金，铂金
玫瑰式切割钻石
雕花祖母绿，凸圆形祖母绿
天然珍珠
内嵌绿松石和螺钿
黑色、象牙色珐琅
2.00 × 10.85 × 5.85 厘米

内设一面镜子，两个翻盖隔层，一个唇膏架。

这款化妆盒的设计灵感源自15～17世纪时波斯书籍的装裱形式。

320

VC 34 A24

Vanity case

Cartier Paris, 1924
Gold, platinum
Rose-cut diamonds
Carved emeralds, emerald cabochons
Natural pearls
Turquoise and mother-of-pearl inlay
Black and ivory-coloured enamel
2.00 × 10.85 × 5.85 cm

The interior fitted with a mirror, two lidded compartments and a lipstick holder.

The design of this piece was inspired by the binding of Persian books dating from the XV to XVII Century.

321

BT 08 A22

臂链

1922 年

卡地亚巴黎为卡地亚伦敦制作，特别订制

铂金

旧式切割钻石

22.3 × 14.0 × 0.20 厘米

此臂链为杜恩吉博伊·伯曼吉爵士特别订制。

臂链挂饰，男女皆可佩戴，是源自印度莫卧儿王朝（16～19世纪）的一种传统手环。这是由卡地亚制作的首款臂链，采用铂金铰链结构，极具柔韧性，与臂部曲线完美契合。镶嵌着钻石的三道圆环（现已遗失），可将臂链固定于手臂。这款首饰也可作为吊坠、胸针、或胸饰佩戴。

杜恩吉博伊·伯曼吉爵士（1862～1937年）来自孟买，是一位船业巨子，在印度和英国均拥有众多地产。他还是一名慈善家，资助印度和英国的众多慈善机构。这款臂链以客人所提供的宝石制作而成。859颗钻石中仅余下28颗（共0.99克拉）未被使用，归还了原主人。

321

BT 08 A22

Bazuband upper arm bracelet

Cartier Paris for London, special order, 1922

Platinum

Old-cut diamonds

22.3 × 14.0 × 0.20 cm

Worn by men as well as women, a bazu band is a traditional Indian bracelet dating from the Mughal dynasty (sixteenth–nineteenth centuries). This piece was the first upper-arm bracelet executed by Cartier. It features an articulated platinum armature that is extremely flexible in order to fit the curve of the arm. Three rings (now missing) set with diamonds made it possible to fasten the bracelet around the arm, although it could also be worn as a pendant, a brooch, or a corsage ornament.

Made to special order for Sir Dhunjibhoy Bomandji.

Sir Parsi Dhunjibhoy Bomandji (1862-1937), originally from Bombay, was a shipping magnate who owned many estates in India and England. As a philanthropist, he supported numerous charities in both India and Britain.

The bracelet was made to order with the client's stones. Of the 859 diamonds supplied, only 28 (for a total of 0.99 carats) were not used and were returned to the client.

322

WP 07 A23

印章挂表

1923 年
卡地亚巴黎
铂金，金
单面切割、玫瑰式和明亮式切割钻石
珍珠
黑玛瑙
黑色珐琅
9.33 × 4.04 厘米

挂表带有一颗祖母绿珠（重17.61
克拉）。表壳嵌有一颗44.15 克拉的雕
花祖母绿。此表为积家126型圆形机
芯，模拟日内瓦波纹装饰，镀铑，8个
调校项目，18颗宝石轴承，瑞士杠杆式
擒纵结构，双金属平衡摆轮，扁平摆轮
游丝。这款表最初搭配的是黑色波纹
丝绸软绳，祖母绿珠作为滑片。

322

WP 07 A23

Seal pendant watch

Cartier Paris, 1923
Platinum, gold
Single-, rose- and brilliant- cut diamonds
Pearls
Onyx
Black enamel
9.33 × 4.04 cm

The pendant with a 17.61-carat emerald bead.

The case with a carved emerald of
44.15 carats.

Round LeCoultre caliber 126
movement, fausses Côtes de Genève
decoration, rhodium-plated, 8 adjustments, 18 jewels, Swiss lever escapement, bimetallic balance, flat balance
spring.

This watch was originally worn on
a black moiré cord, with the emerald
bead functioning as a slider.

棘爪式胸针

1924 年
卡地亚伦敦
铂金, 白金
圆形旧式切割、单面切割钻石
两颗雕花祖母绿（分别重 30.82 克拉和 3.78
克拉）
方形、花式切割和切面圆柱形切割祖母绿
11.0 × 3.90 × 0.90 厘米

323

CL 03 A24

Cliquet brooch

Cartier London, 1924
Platinum, white gold
Round old- and single-cut diamonds
Two carved emeralds (30.82 and 3.78
carats), square, fancy-cut and faceted
calibré-cut emeralds
11.0 × 3.90 × 0.90 cm

324

NE 42 A25

长项链

1925 年
卡地亚纽约，特别订制
铂金
单面切割钻石
一颗六边形雕花祖母绿（重 85.60 克拉）
50 颗刻槽式祖母绿珠（总重约 517 克拉）
天然珍珠
长度　75.05 厘米
吊坠宽度　4.4 厘米

324

NE 42 A25

Sautoir

Cartier New York, special order, 1925
Platinum
Single-cut diamonds
One carved hexagonal 85.60 carat emerald
Fifty fluted emerald beads, weighing an
estimated total of 517 carats
Natural pearls
Length 75.05 cm; width of pendant 4.4 cm

449

325

CL 224 A25

胸针

1925 年
卡地亚巴黎, 特别订制
金, 铂金
旧式切割、玫瑰式切割钻石
枕形红宝石, 切面红宝石圆珠
天然珍珠
无色水晶
黑色珐琅
8.10 × 4.40 厘米

为碧贝斯克公主特别订制。

碧贝斯克公主, 原名玛特·拉沃瓦
(1886～1973年) 来自一个显赫的罗马
尼亚贵族家庭, 同时也是一位知名的
文学家, 在两次世界大战期间主持着
巴黎最时尚的文学沙龙之一。她的座
上宾包括: 马塞尔·普鲁斯特、安那托
尔·弗兰斯、保罗·瓦莱里和保罗·克洛
岱尔。

325

CL 224 A25

Brooch

Cartier Paris, special order, 1925
Gold, platinum
Old- and rose-cut diamonds
Cushion-shaped rubies, faceted ruby
roundels
Natural pearls
Rock crystal
Black enamel
8.10 × 4.40 cm

Made to special order for Princess
Bibesco.

Princess Bibesco (born Marthe La-
hovary, 1886–1973) was a member of
a grand Romanian aristocratic family
as well as a literary figure who hosted
one of the most fashionable literary
salons in Paris during the inter-war
period. Her guests included, among
others, Marcel Proust, Anatole France,
Paul Valéry, and Paul Claudel.

棕榈叶棘爪式扣针

1925 年
卡地亚巴黎
金，铂金
圆形旧式切割、单面切割钻石
面弧形红宝石
雕花穿孔翡翠
黑色珐琅
9.35 × 2.33 厘米

售予威廉·K·范德比尔特夫人（参见CL 92 A05、CL 06 A25、CL 258 A22）。

"在莫卧儿艺术的很多分支中，都可以见到尾端蜷曲的球果形棕榈叶的影子，包括波斯地毯和克什米尔披肩"（参见朱迪·鲁德 ——《卡地亚1900～1939年》第172页）。

翡翠片上有一句中文吉祥语"福在眼前"（来源：《香港佳士得》，1997年11月6日销售目录）。这块翡翠片最早镶嵌在1924年的一个帽针上。

326

CL 244 A25

Boteh cliquet pin

Cartier Paris, 1925
Gold, platinum
Round old- and single-cut diamonds
Buff-top rubies
Carved and pierced jade
Black enamel
9.35 × 2.33 cm

"The palm leaf, or boteh, in the form of a cone shape bent over at the point, is found in many branches of Mughal art, from Persian carpets to Kashmir shawls." (see Judy Rudoe – Cartier 1900-1939, p. 172). The jade plaque depicts a Chinese rebus, Fu zai yan qian, which means "may better luck await you." (source: Christie's Hong Kong, sale catalogue of November 6, 1997). The plaque was initially mounted on a hat pin in 1924. Sold to Mrs. William K. Vanderbilt.

(see CL 92 A05, CL 06 A25, CL 258 A22)

音乐盒

1926 年
卡地亚伦敦
金，铂金
玫瑰式切割钻石
圆柱形红宝石、祖母绿
蓝色珐琅
直径　6.0 厘米

　　盒盖中央为波斯嵌片，上有雌雄鹿及叶形图案，以蓝色珐琅底色，镶嵌钻石和祖母绿。

　　这个音乐盒最早是一个化妆盒，内设一面化妆镜，一个粉盒和一个唇膏架，一个可伸缩抽屉。后来取下了唇膏架，并用音乐装置取代了粉盒。

327

BS 12 A26

Music box

Cartier London, 1926
Gold, platinum
Rose-cut diamonds
Calibrated rubies and emeralds
Blue enamel
Diameter 6.0 cm

The centre with a Persian plaque depicting deer and stag set with diamonds and emeralds on a gold foliate motif and blue enamel ground.

This box was originally a vanity case fitted with a mirror, a powder compartment and a lipstick holder placed in the retractable draw. The powder container was replaced by a music mechanism and the lipstick holder has been withdrawn.

328

CC 87 A26

烟盒

1926 年
卡地亚巴黎为卡地亚伦敦制作
金，铂金
红宝石、蓝宝石、凸圆形祖母绿
玛瑙，青金石
玫瑰式切割钻石
蓝色珐琅
8.75 × 5.63 × 1.81 厘米

盒盖中央为用黄金镌刻式样古典、饰有彩带的圆形图案，边缘镶嵌玫瑰式切割钻石，点缀凸圆形红宝石、祖母绿和蓝宝石，组成一个置身于枝叶中的小鸟图案。

328

CC 87 A26

Cigarette case

Cartier Paris for London, 1926
Gold, platinum
Ruby, sapphire and emerald cabochons
Agate, lapis lazuli
Rose-cut diamonds
Blue enamel
8.75 × 5.63 × 1.81 cm

The centre of the lid applied with an engraved festooned antique gold medallion bordered by rose-cut diamonds and studded with ruby, emerald and sapphire cabochons representing a bird in foliage pattern.

329

WWL 23 A27

手链表

1927 年
卡地亚纽约
铂金
8 颗棒形红宝石，5 个叶形红宝石雕花图案
两颗长阶梯形和一颗玫瑰式切割钻石，圆形旧
式切割钻石和四颗方形钻石
三颗凸圆形宝石
长度　16.0 厘米

此表为积家 103 型椭圆形机芯，模
拟日内瓦波纹装饰，镀铑，8 个调校项
目，19 颗宝石轴承，瑞士杠杆式擒纵结
构，双金属平衡摆轮，宝玑摆轮游丝。

329

WWL 23 A27

Bracelet-watch

Cartier New York, 1927
Platinum
Eight baton-shaped rubies, five ruby motifs
carved in the shape of leaves
Two baguette- and one rose-cut diamonds,
circular-shape old-cut and four square
diamonds
Three cabochon gems
Length 16.0 cm

Oval LeCoultre caliber 103 movement,
fausses Côtes de Genève decoration,
rhodium-plated, 8 adjustments, 19
jewels, Swiss lever escapement, bimetallic balance, Breguet balance spring.

454

330

CC 57 A27

烟盒

1927 年
卡地亚巴黎
黑玛瑙
金, 铂金
玫瑰式切割钻石, 一颗刻槽式红宝石
祖母绿, 红宝石和凸圆形蓝宝石
8.9 × 6.7 × 1.75 厘米

330

CC 57 A27

Cigarette Case

Cartier Paris, 1927
Black onyx
Gold, platinum
Rose-cut diamonds
One fluted ruby
Emeralds, rubies and sapphires cabochons
8.9 × 6.7 × 1.75 cm

印度风格项链和吊坠

1928 年
卡地亚巴黎，特别订制
金
肖像钻石，一颗方形钻石，一颗玫瑰式切割钻石
凸圆形红宝石，雕花红宝石
雕花祖母绿，一颗祖母绿珠
米珠，一颗水滴形珍珠
白玉
彩色珐琅
项链 52.0 厘米（不包括后面的软绳）
吊坠直径 4.2 厘米

这款项链的吊坠为玉石，一面雕花，另一面在玉石上镶嵌雕花红宝石和祖母绿。这块玉石由路易·卡地亚亲手挑选，最早是被用于一条祖母绿珠项链。1928年9月，在一位客人的要求下，用吊坠制作了一条项链。这款项链由"18颗红宝石和钻石嵌片，以及由360颗珍珠串成的5串珠链"组成，具有典型的莫卧儿时期风格，采用了一种被称为"昆丹"镶嵌的独特宝石组合方式，以薄薄的金箔包裹宝石，嵌入金座之中。背面的华丽珐琅则是对莫卧儿人的献礼，正是他们为珐琅技术赋予了高超的艺术地位。佩戴者独享珐琅紧贴肌肤的愉悦。

Indian-style necklace with pendant

Cartier Paris, special order, 1928
Gold
Portrait diamonds, one square diamond,
one rose-cut diamond
Ruby cabochons, engraved rubies
Carved emeralds, one emerald bead
Seed pearls, one drop-shaped pearl
Carved and encrusted white jade
Varicoloured enamels
52.0 cm (necklace without the back cord)
4.2 cm (diameter of the pendant)

Jade can vary in color from milky white to violet to dark green. The pendant on this necklace is of white jade, carved on one side and set with carved rubies and emeralds on the other. It came from a stock of stones personally collected by Louis Cartier, and was originally placed on a necklace of emerald beads.

In September 1928, at the request of a client, the pendant became part of a necklace composed of "eighteen ruby and diamond plaques strung on 360 pearls in five strands," typical of the Mughal period, employing a unique combination of gems in a so-called kundan setting (in which stones are wrapped in thin ribbons of gold against a gold ground). The magnificent enameling on the back is a tribute to the Mughals who elevated enamel techniques to the status of high art. Worn against the skin, this enamel becomes a private pleasure.

332

BT 12 A28

水果锦囊手链

1928 年
卡地亚纽约，特别订制
铂金
明亮式切割、圆形旧式切割和单面切割钻石
叶形雕花红宝石，凸圆形红宝石
镶有钻石底座的刻槽式或光面祖母绿饰珠及蓝
宝石饰珠
黑色珐琅
长度 19.0 厘米

黑色珐琅突出了中央枝干，同时
也制造出一种阴影效果。

332

BT 12 A28

Tutti Frutti strap bracelet

Cartier New York, special order, 1928
Platinum
Brilliant-, round old- and single-cut
diamonds
Leaf-shaped carved rubies, ruby cabochons
Fluted or smooth emerald beads and
sapphire beads studded with collet-set
diamonds
Black enamel
Length 19.0 cm

The black enameling that emphasizes
the central branch creates the effect
of a shadow.

333

CL 98 A28

花瓶胸针

1928 年
卡地亚伦敦
铂金
圆形旧式切割钻石
一颗六边形雕花祖母绿（重 27.14 克拉），
一颗镶钻祖母绿饰珠，
一颗刻槽式祖母绿珠，凸圆形祖母绿
叶形雕花红宝石和凸圆形红宝石
凸圆形蓝宝石
镶钻天然珍珠
3.60 × 4.70 × 1.75 厘米

中间的六边形祖母绿可能是18世
纪～19世纪早期于印度雕刻而成。

333

CL 98 A28

Vase of flowers brooch

Cartier London, 1928
Platinum
Round old-cut diamonds
One hexagonal carved 27.14 carat emerald,
one emerald bead studded with a collet-
set diamond, one fluted emerald bead,
emerald cabochons
Leaf-shaped carved rubies and ruby
cabochons
Sapphire cabochons
Natural pearls studded with collet-set
diamonds
3.60 × 4.70 × 1.75 cm

The hexagonal emerald in the middle
was probably carved in India in the
late eighteenth or early nineteenth
century.

胸针（一对）

1929 年
卡地亚纽约
铂金，14K 白金
明亮式切割、圆形旧式切割和单面切割钻石
叶形雕花红宝石，凸圆形红宝石和红宝石珠
镶有钻石底座的刻槽式或光面祖母绿饰珠，凸
圆形祖母绿
黑色珐琅
4.5×4.1 厘米

这两个胸针最早是一枚椭圆形胸针，改款时加入了一颗叶形红宝石。

334
CL 31 A29

Pair of clip brooches

Cartier New York, 1929
Platinum, 14 carat white gold
Brilliant-, round old- and single-cut
diamonds
Leaf-shaped carved rubies, ruby cabochons
and beads
Fluted or smooth emerald beads studded
with collet-set diamonds, emerald
cabochons
Black enamel
4.5 × 4.1 cm

These two pieces originally formed a
single oval brooch. During the modification, one of the leaf-shaped rubies
was added.

335

EB 05 A29

晚装手袋

1929 年
卡地亚巴黎
金，铂金
一颗雕花红宝石（搭扣）
雕花叶形祖母绿和红宝石
单面切割、长阶梯形切割和方形切割钻石
黑色、白色珐琅
小羊皮
15 × 16 厘米

外框内侧镌刻如下字样：康德·纳斯特夫人，纽约市派克大街1040号。

335

EB 05 A29

Evening bag

Cartier Paris, 1929
Gold, platinum
One engraved ruby (clasp)
Carved leaf-shaped emeralds and rubies
Single-, baguette- and square-cut
diamonds
Black and white enamel
Suede
15 × 16 cm

The frame engraved inside: MRS.
CONDE NAST 1040 PARK AVENUE
NEW YORK CITY.

336

NE 27 A30

长项链

1930 年
卡地亚伦敦，特别订制
铂金
方形明亮式切割、圆形旧式切割、单面切割和玫
瑰式切割钻石
叶形雕花红宝石，凸圆形红宝石和红宝石珠
天然珍珠
总长度　62.05 厘米
大搭扣宽度　2.05 厘米

　　为罗纳德·特里夫人特别订制。

　　这款项饰可拆分为一条手链和一
条短项链。

　　罗纳德·特里夫人生于美国，原名
南希·博金斯（1897～1994年），1945
年嫁给英国保守党议员C·G·兰卡斯
特。在收购著名装饰公司Colefax &
Fowler后，她更名为南希·兰卡斯特，
并成为了一位知名的室内装饰家。

336

NE 27 A30

Sautoir

Cartier London, special order, 1930
Platinum
Square brilliant-, round old-, single- and
rose-cut diamonds
Leaf-shaped carved rubies, ruby cabochons
and beads
Natural pearls
Total length 62.05 cm; width of large clasps
2.05 cm

This sautoir can be broken down into
a bracelet and shorter necklace.

Made to special order for Mrs. Ronald Tree.

Born in the U.S., Mrs. Ronald Tree
(born Nancy Perkins, 1897–1994) married the British Conservative MP C.G.
Lancaster in 1945. After purchasing
the famous decorating firm, Colefax
& Fowler, she became a well-known
interior decorator under the name of
Nancy Lancaster.

337

TB 02 A30

台式烟盒

1930 年
卡地亚纽约
金
白色、黄色、绿色、蓝色珐琅
红漆
14.50 × 8.80 × 5.20 厘米

盒盖中央有一幅沙·贾汗的细密
画。盒盖内侧带黄金嵌片，黑色珐琅铭
文：沙·贾汗，1628～1658年在位，阿
格拉泰姬陵建造者。

337

TB 02 A30

Cigarette table box

Cartier New York, 1930
Gold
White, yellow, green and blue enamel
Red lacquer
14.50 × 8.80 × 5.20 cm

The cover applied to the centre with
a miniature depicting Shah Jahan.
The interior of the cover with a gold
plaque and a black enamel inscription reading: SHAH JAHAN - REIGNED
1628-1658 BUILDER OF THE TAJ MA-
HAL AT AGRA.

338

PB 27 A31-33

袖珍粉盒和唇膏盒

1931 年和 1933 年

卡地亚巴黎

金, 铂金

凸圆形红宝石, 单面切割和长阶梯形切割钻石

黑色珐琅

袖珍粉盒　4.70 × 3.70 × 1.50 厘米

唇膏盒　4.70 × 1.80 × 1.50 厘米

338

PB 27 A31-33

Miniature powder compact and matching lipstick holder

Cartier Paris, 1931 and 1933

Gold, platinum

Ruby cabochons, single- and baguette-cut
diamonds

Black enamel

Powder compact 4.70 × 3.70 × 1.50 cm

Lipstick holder 4.70 × 1.80 × 1.50 cm

339

NE 25 A32

项链

1932 年
卡地亚伦敦，特别订制
铂金
圆形旧式切割和玫瑰式切割钻石
一颗枕形抛光祖母绿（重143.23 克拉）
中央高度　8.80 厘米

为格兰纳德夫人特别订制。

格兰纳德夫人（原名碧翠斯·米尔斯）是美国金融家和慈善家奥格登·米尔斯之女。她于1909年嫁给格兰伯格伯爵八世。作为卡地亚伦敦的常客，她尤其钟情于卡柯史尼克冠冕，在1922到1937年间一共订制了三顶。

339

NE 25 A32

Necklace

Cartier London, special order, 1932
Platinum
Round old- and rose-cut diamonds
One cushion-shaped polished 143.23 carat emerald
Height at centre 8.80 cm

Made to special order for Lady Granard.

Lady Granard (born Beatrice Mills) was the daughter of the American financier and philanthropist Ogden Mills. She married the eighth Earl Granard in 1909. As a regular client of Cartier London, she was particularly fond of kokoshnik tiaras, ordering three between 1922 and 1937.

340

EG 10 A35

耳坠（一对）

1935 年
卡地亚巴黎，特别订制
铂金，白金
旧式切割、单面切割钻石
切面和圆柱形切割红宝石
天然珍珠
5.50 × 1.20 厘米

为印度巴提亚拉土邦王公布品德
拉·塞恩勋爵特别订制。

以客人所提供的一对耳坠制成，
完美地体现了卡地亚对传统印度饰品
的诠释。

340

EG 10 A35

Pair of ear-pendants

Cartier Paris, special order, 1935
Platinum, white gold
Old- and single-cut diamonds
Faceted and calibré-cut rubies
Natural pearls
5.50 × 1.20 cm

Made from a pair of pendant earrings
supplied by the client, these earrings
are a good example of Cartier's interpretation of traditional Indian jewellery.

Made to special order for Sir
Bhupindra Singh, Maharajah of Patiala.

34I

NE 28 A36

水果锦囊项链

1936 年制作，1963 年改款
卡地亚巴黎，特别订制
铂金，白金
马眼形、长阶梯形和圆形旧式切割钻石
13 颗明亮式切割蓝宝石（总重 146.9 克拉），两
颗叶形雕花蓝宝石（分别重 50.80 克拉和 42.45
克拉），蓝宝石珠，一颗凸圆形蓝宝石
方形雕花祖母绿，刻槽式和光面祖母绿珠，凸圆
形祖母绿
叶形雕花红宝石，光面雕花红宝石珠，凸圆形红
宝石
长度　43.00 厘米（打开后）

为黛丝·法罗斯太太特别订制，后
传给其女儿卡斯特加伯爵夫人（参见
BT 44 A21、VC 85 A13、EG 28 A63）。

这款项链最初为印度风格，搭配
黑色软绳，在背后固定。后经客人要求
进行了重新镶嵌。

34I

NE 28 A36

Tutti Frutti necklace

Cartier Paris, special order, 1936, altered in
1963
Platinum, white gold
Marquise-, baguette- and round old-cut
diamonds
Thirteen briolette-cut sapphires weighing
146.9 carat in total, two leaf-shaped carved
sapphires (50.80 and 42.45 carats), sapphire
beads, one sapphire cabochon
Square carved emeralds, fluted and smooth
emerald beads, emerald cabochons
Leaf-shaped carved rubies, smooth and
engraved ruby beads, ruby cabochons
Length 43.00 cm (open)

Originally made in the Indian style,
with black cord fastening round the
back, this necklace was remounted
upon the request of the client.

Made to special order for Mrs. Daisy Fellowes who passed it on to the
Countess of Castéja (her daughter).

(see BT 44 A21, VC 85 A13, EG 28
A63)

长阶梯形腕表式胸针

1938 年
卡地亚巴黎
铂金，金
一颗五边形雕花祖母绿
长阶梯形切割、单面切割钻石
2.64 × 2.58 × 1.10 厘米

表为积家104型长方形机芯，日内瓦波纹装饰，镀铑，2个调校项目，17颗宝石轴承，瑞士杠杆式擒纵结构，双金属平衡摆轮，扁平摆轮游丝。

这款表最早搭配了一条彩色（水果锦囊）手链。

342

WB 26 A38

Baguette watch-brooch

Cartier Paris, 1938
Platinum, gold
A 5-sided carved emerald
Baguette- and single-cut diamonds
2.64 × 2.58 × 1.10 cm

Rectangular LeCoultre caliber 104 Duoplan movement, Côtes de Genève decoration, rhodium-plated, 2 adjustments, 17 jewels, Swiss lever escapement, bimetallic balance, flat balance spring.

This watch was originally mounted on a colorful multigem (*Tutti Frutti*) bracelet.

Winding crown on back.

343

CL 58 A38

印度风格胸针

1938 年
卡地亚巴黎
金
肖像钻石
切面蓝宝石
红宝石和凸圆形祖母绿
圆形和巴洛克天然珍珠
凸圆形绿松石和绿松石珠
红色、绿色珐琅
10.7 × 10.8 厘米

这款胸针是卡地亚对印度珠宝的有趣诠释，以档案中所描述的古老元素制作而成："4种印度图案……由11颗古老的金色橄榄组成的5条串珠。"

343

CL 58 A38

Indian-style clip brooch

Cartier Paris, 1938
Gold
Portrait diamonds
Faceted sapphires
Ruby and emerald cabochons
Round and baroque natural pearls
Turquoise cabochons and beads
Red and green enamel
10.7 × 10.8 cm

As an interesting example of Cartier's interpretation of Indian jewellery, this brooch was made from old elements described in the archives as a "cluster of four Hindu motifs . . . with five strands of eleven old gold olives."

344

BT 51 A39

印度风格手链

1939 年
卡地亚伦敦，特别订制
黄金、白金、铂金
明亮式切割、单面切割钻石
天然珍珠
19.0 × 3.7 厘米

344

BT 51 A39

Indian-style bracelet

Cartier London, special order, 1939
Yellow gold, white gold, platinum
Brilliant- and single-cut diamonds
Natural pearls
19.0 × 3.7 cm

345

CL 241 A39

印度风格胸针（一对）

1939 年
卡地亚伦敦
金
肖像钻石
红宝石和凸圆形祖母绿
天然珍珠
3.2 × 3.3 厘米

卡地亚在两个古老的印度图样的基础上，增加了26颗珍珠和夹扣，制作出一对胸针。

345

CL 241 A39

Indian-style pair of clip brooches

Cartier London, 1939
Gold
Portrait diamonds
Ruby and emerald cabochons
Natural pearls
3.2 × 3.3 cm

Cartier added twenty-six pearls and clasps to two old Indian motifs in order to produce this pair of brooches.

346

EG 28 A63

耳坠（一对）

1963 年
卡地亚巴黎，特别订制
铂金，白金，黄金
明亮式切割、圆形旧式切割、单面切割和玫瑰式
切割钻石
镶钻雕花祖母绿和祖母绿饰珠
黑色珐琅
5.20 × 1.89 厘米

来源于黛丝·法罗斯夫人和卡斯
特加伯爵夫人。

两颗雕花祖母绿是从一枚胸针上
取下，由黛丝·法罗斯夫人于1939年带
至卡地亚。这两颗祖母绿最早是镶嵌
在一个帽针上，后于1945年由卡地亚
伦敦重新改款，与钻石珠一起，作为一
款珍珠项链的搭扣。法罗斯夫人的女
儿卡斯特加伯爵夫人于1963年请卡地
亚巴黎将其改为耳环，与其母亲的"印
度"项链搭配（参见BT 44 A21，VC 85
A13，NE 28 A36）。

346

EG 28 A63

Pair of ear-pendants

Cartier Paris, special order, 1963
Platinum, white gold, yellow gold
Brilliant-, round old-, single- and rose-cut
diamonds
Carved emeralds and emerald beads
studded with collet-set diamonds
Black enamel
5.20 × 1.89 cm

Provenance: Mrs. Daisy Fellowes and
the Comtesse de Castéja

The two carved emeralds were
taken from a brooch brought to Cartier by Daisy Fellowes in 1939. First set
on a hat pin, they were used again in
1945 by Cartier London, in conjunc-
tion with diamond-studded beads, to
decorate the clasp of a pearl necklace.
Fellowes's daughter, the Comtesse de
Castéja, ultimately had Cartier Paris
turn them into earrings in 1963, designed to be worn with her mother's
"Hindu" necklace.

(see BT 44 A21, VC 85 A13, NE 28
A36)

1847

路易·弗朗索瓦·卡地亚(1819～1904年)从其师傅阿道夫·皮卡手中接管下位于巴黎蒙特吉尔街29号的珠宝店。

1856

拿破仑一世的侄女,法国皇帝拿破仑三世的堂妹,玛蒂尔德公主首次在卡地亚订制珠宝。

1859

卡地亚搬迁至当时巴黎的时尚中心——意大利大道9号。同年,欧珍妮皇后成为卡地亚的客户。

1888

卡地亚首款女士珠宝链表面世。

1899

卡地亚位于巴黎黄金地段的和平街13号精品店开幕。阿尔弗雷德(1841～1925年)的长子路易·卡地亚(1875～1942年)投身家族事业,与父亲共事一年。

1900

各国皇家成员和贵族竞相购买灵感源自18世纪,镶嵌于铂金基座上的新古典风格钻饰。

1902

阿尔弗雷德次子皮埃尔·卡地亚(1878～1964年)于英国国王爱德华七世加冕之际在伦敦新博灵顿大街4号开设分店。

1904

卡地亚获得第一份皇家委任状,成为英国国王爱德华七世的御用供应商。

同年,获得西班牙国王阿方索十三世颁发的皇家委任状。

路易·卡地亚为友人巴西飞行员阿尔伯特·山度士·杜蒙创制了一款装配皮表带的腕表,可佩戴于手腕之上。

英国亚历珊德拉王后购买了一条充满印度风情的项链。

第一批带有"装饰艺术"风格的珠宝问世,以抽象和个性的几何造型为其主要特色。

1905

获得葡萄牙国王卡洛斯一世颁发的皇家委任状。

1906

阿尔弗雷德·卡地亚的第三子雅克执掌伦敦分店。

推出首款Tonneau酒桶形腕表。

1907

卡地亚在圣彼得堡欧洲大酒店举行首次展卖。

1908

暹罗国王帕拉米达·马哈·朱拉隆功为卡地亚颁发皇家委任状。

1909

卡地亚在伦敦新庞德街175～176号设立新店。

皮埃尔·卡地亚在纽约第五大道712号设立分店。

1910

皮埃尔·卡地亚将蓝色的"希望"之钻售予美国客户艾弗琳·沃尔什。

1911

卡地亚推出Santos腕表,设计灵感源自1904年的表款。

雅克·卡地亚(1884～1942年)到波斯湾及印度旅行,觐见印度王室。

1912

第一款A型魅幻时钟问世。

巴黎市政厅将卡地亚制作的皇家彩蛋作为官方礼品赠送给沙皇尼古拉二世,这份礼物现收藏于纽约大都会博物馆。

同年首次推出长阶梯形切割的钻石和彗星时钟。

1913

卡地亚获得塞尔维亚国王彼得一世颁发的皇家委任状。

1914

在女装圆形腕表的表圈铺镶钻石和黑玛瑙,创造出第一个"豹纹"图案。

获得奥尔良菲利普公爵颁发的皇家委任状。

1917

纽约店铺迁至第五大道653号。此处宅邸原为莫顿·普朗特所有，由皮埃尔·卡地亚以一条由55颗和73颗天然珍珠构成的双串式珍珠项链交换而得。

第一幅Tank坦克腕表的设计样图问世。

创办卡地亚纽约工作室，后来命名为美国艺术工作坊。

1918

为马歇尔将军和贝当设计司令权杖。

1919

推出Tank坦克腕表。

获得比利时国王阿尔伯特一世颁发的皇家委任状。

1921

获得英国王储威尔士亲王（即未来的英国国王爱德华八世，后于1936年逊位，成为温莎公爵）颁发的皇家委任状。

1922

设立伦敦卡地亚工作室，并命名为"英国艺术工作坊"。

1923

创制首款Portique门廊魅幻时钟。

1924

推出三环戒指和手镯，以三色金制成，在美国被称为Trinity戒指。

让·考克多佩戴这枚戒指，使其成为巴黎社交界争相追捧的时尚之作。

1925

卡地亚出席在巴黎"优雅之亭"举行的"国际现代装饰与工艺艺术展览"。

1926

卡地亚珠宝及其著名红盒子亮相百老汇舞台，出现在安尼塔·卢斯主演的戏剧《绅士爱美人》之中。

1928

美国忠实顾客马乔里·梅瑞威瑟·波斯特于卡地亚伦敦精品店购得一对曾为法国玛丽·安东尼王后所佩戴的耳坠。

创制Tortue龟形单钮计时腕表。

1929

获得埃及国王福阿德一世颁发的皇家委任状，参加在开罗举办的法国艺术展。

1933

贞·杜桑被任命为卡地亚高级珠宝总监。

卡地亚获得"隐形镶嵌"的发明专利。这是一种宝石镶嵌技术，可隐藏金属基座，只显露出宝石。

1935

卡地亚蒙特卡罗分店开幕。

1938

卡地亚戛纳分店开幕。

卡地亚制作出全球最小腕表，并将之献给英国伊丽莎白公主。

1939

获得阿尔巴尼亚国王佐格一世颁发的皇家委任状。

1940

卡地亚资助戴高乐将军在伦敦组织的自由法国运动。

戴高乐将军的部份演讲更是在雅克·卡地亚特别为将军所准备的办公室内所撰写。

1942

创制"笼中鸟"胸针，象征被德国占领下的法国。法国1944年解放之时，又设计出"自由鸟"胸针作为纪念。

1945

皮埃尔·卡地亚接管卡地亚巴黎的业务。

路易之子克劳德接管卡地亚纽约业务。

雅克之子让·雅克·卡地亚接管卡地亚伦敦业务。

1947

卡地亚百年庆典。

1949
温莎公爵夫妇于巴黎购得一枚铂金豹形胸针，镶嵌着一颗152.35克拉的凸圆形克什米尔蓝宝石。1987年，卡地亚购回这枚胸针，收藏于卡地亚艺术典藏系列。

1950
好莱坞传奇女星格洛利亚·斯旺森佩戴1930年自卡地亚购得的两只钻石和无色水晶手镯亮相于电影《日落大道》。

1953
玛丽莲·梦露在电影版《绅士爱美人》中演唱歌曲《卡地亚！》。

1955
卡地亚为让·考克多制作了一把由他自己设计的法兰西学院院士佩剑。

1956
格蕾丝王妃与兰尼埃亲王喜结连理，获赠大量由卡地亚制作的珠宝礼品，包括她的订婚戒指，镶嵌一颗重12克拉的祖母绿式切割钻石。

1957
卡地亚忠实客户芭芭拉·赫顿购得一枚以黄金、黑玛瑙和黄钻制成的虎形胸针。

1968
墨西哥女演员玛莉亚·菲利克斯委托卡地亚制作了一条蛇形钻石项链。

1969
卡地亚购入一颗重69.42克拉的梨形钻石，将其售于理查·波顿，后者将其作为生日礼物赠予伊莉萨白·泰勒。这颗著名的卡地亚钻石后来被命名为泰勒·波顿之钻。卡地亚"Love手镯"诞生。

1974
卡地亚将大量装饰艺术风格珠宝出借给《伟大的盖茨比》剧组。该片由杰克·克莱顿执导，罗伯特·雷德福和米亚·法罗主演。

1975
卡地亚庆祝路易·卡地亚百年诞辰。
第一个大型回顾展于蒙特卡洛开幕。

1979
卡地亚巴黎、卡地亚伦敦、卡地亚纽约重新合并成为一个法律主体。

1983
创建卡地亚古董收藏室（后来的卡地亚艺术典藏室），记录和阐述卡地亚的艺术和历史演变。

1984
在儒伊·昂·若塞斯创建卡地亚当代艺术基金会。

1989~1990
首次在巴黎小皇宫举办"卡地亚艺术典藏系列"大型回顾展。

1992
在圣彼得堡艾尔米塔什博物馆举办"卡地亚艺术典藏展"。

1994
卡地亚当代艺术基金会迁至位于巴黎左岸的拉斯巴耶大道，由让·努维尔设计的一幢建筑内。

1995
在日本东京都庭园美术馆举办"卡地亚艺术典藏，法国珠宝艺术世界"展览。

1996
在瑞士洛桑遗产基金会举行回顾展"卡地亚：珠宝之光"。

1997
卡地亚庆祝150周年华诞，在伦敦大英博物馆和纽约大都会博物馆举办"卡地亚：1900～1939年"回顾展。

1999
在墨西哥城美术宫举办"卡地亚艺术典藏，时代之辉煌"回顾展，并有幸邀请到墨西哥著名女演员玛莉亚·菲利克斯担当嘉宾。
在芝加哥富地自然历史博物馆举办"卡地亚：1900～1939年"回顾展。

2002

在柏林维特拉设计博物馆和米兰王宫举办名为"索特萨斯眼中的卡地亚设计"的展览。

后又于东京醍醐寺和休斯顿美术馆举办同名巡回展。

2003

"金伯利进程"开始实施。卡地亚迅速执行相关措施，结束冲突钻石交易。

卡地亚资助在巴黎蓬皮杜中心举办名为"让·考克多：世纪华彩"的展览。

2004

卡地亚在上海博物馆举办"卡地亚艺术珍宝展"。

2005

卡地亚成为世界女性论坛成员，为其提供支持。

2006

卡地亚于世界妇女论坛为女性创业家颁发特别奖。

2007

在里斯本卡洛斯特·古尔本金安基金会博物馆举办"卡地亚1899~1949年：风格之旅"。在莫斯科克里姆林宫博物馆举办名为"卡地亚：20世纪创新旅程"的展览。

2008

在韩国首尔德寿宫国立博物馆举办"卡地亚艺术典藏"回顾展。

2009

在日本东京国立博物馆举办名为"卡地亚：与美丽相逢的记忆"的展览，由日本设计师吉冈德仁策展。

CHRONOLOGY

1847

Louis-François Cartier (1819-1904) took over the jewellery workshop of his apprenticeship master Adolphe Picard at 29, Rue Montorgueil in Paris.

1856

Princess Mathilde, niece of Napoleon I and cousin to Emperor Napoleon III, made her first purchase from Cartier.

1859

Cartier moved to 9, Boulevard des Italiens. Empress Eugénie became a customer.

1888

The first jewellery bracelet-watches for ladies.

1899

Cartier opened at 13, Rue de la Paix in Paris. Louis Cartier (1875-1942), eldest son of Alfred (1841-1925), had been in business with his father for one year.

1900

Crowned heads and aristocrats from around the world flocked to buy neoclassic diamond jewellery mounted in platinum of Eighteenth Century inspiration.

1902

Pierre Cartier (1878-1964), Alfred's second son, opened a branch at 4, New Burlington Street in London. The opening coincided with the coronation of King Edward VII.

1904

Cartier received its first royal warrant as official purveyor to King Edward VII of England.
Appointment as official purveyor to King Alfonso XIII of Spain.
Louis Cartier created a wristwatch with a leather strap, expressly designed to be worn on the wrist, for his friend the Brazilian aviator Alberto Santos-Dumont.
Queen Alexandra of England purchased a necklace in the Indian style.
The first jewellery prefiguring the Art Deco style, recognisable for its abstract and geometric forms.

1905

Appointment as official purveyor to King Carlos I of Portugal.

1906

Alfred Cartier's third son, Jacques, took over the London branch.
Creation of the first Tonneau wristwatch.

1907

First exhibition and sale in Saint Petersburg, at the Grand Hotel Europe.
Appointment as official purveyor to Tsar Nicholas II of Russia.

1908

Appointment as official purveyor to King Paramindr Maha Chulalongkorn of Siam.

1909

Opening of a new address at 175-176, New Bond Street, London.
Pierre Cartier opened a subsidiary in New York at 712, Fifth Avenue.

1910

Pierre Cartier sold the blue Hope Diamond to an American customer, Evalyn Walsh McLean.

1911

Launch of the Santos de Cartier wristwatch, inspired by the 1904 model.

Jacques Cartier (1884-1942) travelled to India to attend the Delhi Durbar and to the Persian Gulf.

1912

The first mystery clock: Model A.

A delegation from the City of Paris presented Tsar Nicholas II with the Cartier Imperial Egg (now at the Metropolitan Museum of Art in New York).

First baguette-cut diamonds.

First Comet clocks.

1913

Appointment as official purveyor to King Peter I of Serbia.

1914

The bezel on a lady's round wristwatch was paved with diamonds and onyx to create the first "panther" motif.

Appointment as official purveyor to Duke Philippe of Orleans.

1917

The New York store moved to 653, Fifth Avenue, the mansion home of Morton F. Plant which Pierre Cartier acquired in exchange for a double-strand necklace of 55 and 73 natural pearls.

First studies for the Tank wristwatch.

Opening of the Cartier New York workshop later named "American Art Works".

1918

Creation of batons for Field-Marshals Foch and Pétain.

1919

Launch of the Tank wristwatch.

Appointment as official purveyor to King Albert I of Belgium.

1921

Appointment as official purveyor to the Prince of Wales, future King Edward VIII who, on abdicating in 1936, became the Duke of Windsor.

1922

Opening of Cartier London workshop named "English Art works".

1923

The first Portique mystery clock.

1924

Creation of the three-band ring and bracelet combining gold in three colours, known in the United States as Trinity.

Jean Cocteau adopted the ring and made it fashionable among Parisian society.

1925

Cartier made a memorable appearance at the Exposition internationale des Arts décoratifs et industriels modernes in Paris, in the Pavillon de l'Élégance.

1926

Cartier jewellery in its famous red box appeared on the Broadway stage in Anita Loos' play "Gentlemen Prefer Blondes".

1928

A loyal American customer, Marjorie Merriweather Post, purchased from Cartier in London earrings once worn by Queen Marie-Antoinette of France.

Creation of the Tortue single button chronograph wristwatch.

1929

Appointment as official purveyor to King Fouad I of Egypt and participation in the Exhibition of French Arts in Cairo.

1933

Jeanne Toussaint was made head of Cartier Fine Jewellery.

Cartier filed a patent for the "invisible mount", a stone-setting technique in which the metal of the mount disappears to show only the stones.

1935

Cartier opened in Monte-Carlo.

1938

Cartier opened in Cannes.

The smallest wristwatch in the world, by Cartier, was given to Princess Elizabeth of England.

1939

Appointment as official purveyor to King Zog I of Albania.

1940

General de Gaulle founded the Forces Française Libres movement in London, for which he received Cartier's steadfast support. Some of his speeches were written in the office which Jacques Cartier placed at the general's disposal.

1942

Creation of the "Caged Bird" brooch as a symbol of the Occupation. In 1944, the "Freed Bird" brooch celebrated the Liberation of Paris.

1945

Pierre Cartier was now at the head of Cartier Paris. Claude, son of Louis, took the helm of Cartier New York while Jean-Jacques Cartier, son of Jacques, was at the head of Cartier London.

1947

Cartier celebrated its centenary.

1949

The Duke and Duchess of Windsor purchased, in Paris, a platinum panther brooch on a Kashmir cabochon sapphire of 152.35 carats. Cartier would buy the brooch for its own collection in 1987.

1950

The legendary Hollywood actress Gloria Swanson appeared in "Sunset Boulevard" wearing the two diamond and rock crystal bracelets she had bought from Cartier in 1930.

1953

Marilyn Monroe sang "Cartier!" in the film version of "Gentlemen Prefer Blondes".

1955

Creation of Jean Cocteau's sword for his election to the Académie Française, from the artist's own design.

1956

For her marriage to Prince Rainier, Princess Grace received numerous gifts of jewellery by Cartier including her engagement ring, set with an emerald-cut diamond of 12 carats.

1957

A loyal customer, Barbara Hutton purchased a tiger brooch in yellow gold, onyx and yellow diamonds.

1968

The Mexican actress María Félix commissioned Cartier to make a diamond necklace in the form of a serpent.

1969

Cartier acquired an exceptional pear-shaped diamond of 69.42 carats and sold it to Richard Burton. He gave it to Elizabeth Taylor for her birthday. The famous Cartier Diamond was thus renamed the Taylor-Burton.

Creation of the "Love bracelet".

1974

Cartier loaned a significant part of its Art Deco style jewellery collection for "The Great Gatsby", directed by Jack Clayton and starring Robert Redford and Mia Farrow.

1975

Cartier celebrated the one-hundredth anniversary of the birth of Louis Cartier.

Opening in Monte-Carlo of the first major retrospective.

1979

Cartier Paris, Cartier London and Cartier New York were reunited as a single legal entity.

1983

Creation of the Collection Ancienne Cartier (later the Cartier Collection) to record and illustrate how the jeweller's art and history have evolved.

1984

Creation of the Fondation Cartier pour l'art contemporain in Jouy-en-Josas.

1989-1990

"The Art of Cartier", the first major retrospective in Paris, at the Petit Palais.

1992

"The Art of Cartier" exhibition at the Hermitage Museum in Saint Petersburg.

1994

The Fondation Cartier pour l'art contemporain moved to the Left Bank in Paris in a building on Boulevard Raspail, the work of the architect Jean Nouvel.

1995

"The Art of Cartier, the World of French Jewellery Art" exhibition at the Tokyo Metropolitan Teien Art Museum in Japan.

1996

"Cartier, Splendours of Jewellery", a retrospective exhibition at the Hermitage Foundation in Lausanne, Switzerland.

1997

Cartier celebrated its 150th anniversary : "Cartier 1900-1939" retrospective at the British Museum in London and the Metropolitan Museum of Art in New York.

1999

"The Art of Cartier, A splendor of Time" retrospective at the Museo del Palacio de Bellas Artes in Mexico City with guest of honour the Mexican actress María Félix.

"Cartier 1900-1939" at the Field Museum of Natural History in Chicago.

2002

"Cartier Design viewed by Ettore Sottsass" at the Vitra Design Museum in Berlin and the Palazzo Reale in Milan.

The exhibition would later travel to the Daigoji Temple in Kyoto and the Houston Museum of Fine Arts.

2003

The Kimberley Process was implemented and Cartier immediately adopted measures to end trade in conflict diamonds.

Cartier contributed to "Jean Cocteau, Spanning the Century" at the Centre Georges Pompidou in Paris.

2004

Cartier presented "The Art of Cartier", a retrospective exhibition at the Shanghai Museum.

2005

Cartier gave its support to the Women's Forum by becoming an active member.

2006

Cartier presented a special award for women business entrepreneurs at the Women's Forum.

2007

"Cartier 1899-1949, The Journey of a Style" at the Calouste Gulbenkian Foundation Museum in Lisbon.
"Cartier, Innovation through the 20th Century" at the Kremlin Museum in Moscow.

2008

"The Art of Cartier", a retrospective exhibition held at the Deoksugung National Museum of Seoul, Korea.

2009

"Story of…, Memories of Cartier creations" directed by the Japanese designer Tokujin Yoshioka, held at the Tokyo National Museum, Japan.

BIBLIOGRAPHY

1 - Cartier

GAUTIER, Gilberte. Rue de la Paix. Julliard, Paris 1980.

NADELHOFFER, Hans. Cartier, Éditions du Regard. Paris, 2007. French version
NADELHOFFER, Hans. Cartier, Thames & Hudson. Londres, 2007. English version
NADELHOFFER, Hans. Cartier, Chronicle Books. New York. 2007. American version
NADELHOFFER, Hans. Cartier, Federico Motta Editore , Milan. 2007. Italian version
NADELHOFFER, Hans. Cartier, Editions du Regard. Paris, 1984.
NADELHOFFER, Hans. Cartier, Jewellers Extraordinary. Thames & Hudson, Londres 1984.
NADELHOFFER, Hans. Cartier, Jewellers Extraordinary. Harry N. Abrams, Inc Publishers, New York. 1984.
NADELHOFFER, Hans. Cartier, Longanesi, Milan.
NADELHOFFER, Hans. Cartier, Bijutsu Shuppan Sha Co, Ltd. Tokyo,
NADELHOFFER, Hans. Cartier – Juwelier der Könige, Könige der Juweliere.
Herrsching am Ammersee, Schuler Verlagsgesellschaft. Allemagne.

GAUTIER, Gilberte. La Saga des Cartier 1847-1988. Paris, Michel Lafon.1988
Updated version of Rue de la Paix.
GAUTIER, Gilberte. Cartier the Legend. Arlington Books Ltd, Londres.
GAUTIER, Gilberte La Saga dei Cartier. Sperling & Kupfer Editori, Milan.

BARRACA, J., NEGRETTI G., NENCINI, G. Le Temps de Cartier. Wrist International S.r.l, Paris, 1989.
BARRACA, J., NEGRETTI G., NENCINI, G. Le Temps de Cartier. 1st édition, Wrist International S.r.l, Milan.
BARRACA, J., NEGRETTI G., NENCINI, G. Le Temps de Cartier. 2nd édition, Publi Prom, Milan, 1993
BARRACA, J., NEGRETTI G., NENCINI G. Le Temps de Cartier. English version : Wrist International S.r.l, Milan

COLOGNI, F. MOCCHETTI, E. L'Objet Cartier : 150 ans de tradition et d'innovation
La Bibliothèque des Arts, Paris, Lausanne,1992
COLOGNI, F. MOCCHETTI, E. L'oggetto Cartier : 150 anni di tradizione e innovazione
Giorgio Mondadori, Milan 1993.
COLOGNI, F. MOCCHETTI, E. Made by Cartier: 150 years of Tradition and Innovation
Abbeville Press, New York 1993.
COLOGNI, F. MOCCHETTI, E. Creador por Cartier : 150 años de tradición y Inovación
Giorgio Mondadori, Milan 1993.

COLOGNI, F. NUSSBAUM, E. Cartier, le joaillier du platine
La Bibliothèque des Arts, Paris, Lausanne,1995
COLOGNI, F. NUSSBAUM, E. Platinum by Cartier, The Triumphs of the Jewellers' Art
Harry N. Abrams, Inc., Publishers, New York.
COLOGNI, F. NUSSBAUM, E. Cartier Meisterwerke aus Platin
Herrsching am Ammersee, Bruckman Verlag

COLOGNI, F. NUSSBAUM, E. Cartier L'arte del platino
Editoriale Giorgio Mondadori, Milan.
COLOGNI, F. NUSSBAUM, E. Cartier le Joaillier du platine
EDICOM, Tokyo

TRETIACK, Philippe, Cartier
Editions Assouline, coll. "La mémoire des marques",
Paris, 1996
TRETIACK, Philippe, Cartier
Thames & Hudson, Londres, 1996
TRETIACK, Philippe, Cartier
Universe Publishing / The Vendome Press, New York,
1997
TRETIACK, Philippe, Cartier
Schirmer / Mosel, Munich, 1997
TRETIACK, Philippe, Cartier
Korinsha Press, Kyoto, 1997

COLOGNI, Franco, Cartier La montre Tank
Flammarion, Paris, 1998
COLOGNI, Franco, Cartier The Tank watch
Flammarion, Paris
COLOGNI, Franco, Cartier Die Tank Uhr
Flammarion, Paris
COLOGNI, Franco, Cartier L'Orologio Tank
Flammarion, Paris
COLOGNI, Franco, Cartier El Reloj Tank
Flammarion, Paris
COLOGNI, Franco, Cartier la montre Tank
Versions japonaise et chinois traditionnel
Flammarion, Paris

CHAILLE, François, Cartier : Styles et stylos
Flammarion, Paris 2000
CHAILLE, François, Creative Writing
Flammarion, Paris
CHAILLE, François, le Penne di Cartier
Flammarion, Paris

Les Must de Cartier
Editions Assouline, Paris, 2002
Les Must de Cartier
English, Italian and Japanese versions
Editions Assouline, Paris, 2002

ALVAREZ, José, Cartier l'Album
Editions du Regard, Paris, 2003

COLOGNI, Franco, NUSSBAUM, Eric, CHAILLE, François
La Collection Cartier Tome 1 , la Joaillerie
Editions Flammarion, Paris, 2004

COLOGNI, Franco, CHAILLE, François
La Collection Cartier Tome 2, l'Horlogerie
Editions Flammarion, Paris, 2006

COLENO, Nadine. Étourdissant Cartier, la création
depuis 1937, published by Éditions du Regard, on sale
from September 2008. French version

COLENO, Nadine. Amazing Cartier, creations since
1937, published by Flammarion, on sale in the United
Kingdom and United States from April 2009. English
version

2 – Exhibition catalogues of the Cartier Collection

NADELHOFFER, Hans, Retrospective Louis Cartier : One Hundred and one Years of the Jeweller's Art. Cartier Inc., New York, 1976.

NADELHOFFER, Hans, Retrospective Louis Cartier : One Hundred and one Years of the Jeweller's Art. Exhibition at the Coundy Museum, Cartier Inc., Los Angeles, 1982.

BUROLLET, Thérèse, CHAZAL, Yves, PIVER-SOYEZ, Sylvie-Jan
Exhibition catalogue « L'Art de Cartier », at Musée du Petit Palais in Paris and also at Accademia Valentino, Rome. Paris-Musées, Paris. 1989.
Italian version published by Muse, Bologne.
English version The Art of Cartier Paris-Musées, Paris.

CHAZAL, Yves, SOUSLOV, V.
Exhibition catalogue « L'Art de Cartier », musée de l'Ermitage, Saint-Pétersbourg.
Les Éditions du Mécène, Paris. 1992.

PERRIN, Alain Dominique, NUSSBAUM, Eric, CHAZAL, Martine, UNNO, Hiroshi, TAKANAMI, Machiko.
Exhibition catalogue « L'Art de Cartier », Metropolitan Teien Art Museum, Tokyo.
Anglo-japanese version published by Nihon Keizai Shimbun, Inc., Tokyo. 1995.

DAULTE, François, NUSSBAUM, Eric
Exhibition catalogue « Cartier, Splendeurs de la joaillerie », Fondation de l'Hermitage, Bibliothèque des Arts, Lausanne (1996).

RUDOE, Judie
Exhibition catalogue « Cartier 1900-1939 », at Metropolitan Museum of Art – New York, British Museum - Londres (1997) and at Field Museum – Chicago (1999-2000).
The British Museum Press, Londres, 1997 .
N. Abrams Inc.Publishers, The Metropolitan Museum of Art, New York, 1997.

TOVAL, Rafael, ESTRADA, Geraldo, NUSSBAUM, Eric, MONSIVAIS, Carlos, ARTEAGA, Agustin, ALFARO, Alfonso.
Exhibition catalogue « El Arte de Cartier - Resplandor del Tiempo » at l'Instituto Nacional de Bellas Artes, Mexico City. Americo Arte Editores, Mexico, 1999.

COLOGNI, Franco, SOTTSASS, Ettore, JOUSSET, Marie-Laure, NUSSBAUM, Eric, KRIES, Mateo, VON VEGESACK, Alexander, JAIS, Betty, KARACHI, Jacqueline
« Cartier Design – Eine Inszenierung von Ettore Sottsass" at Vitra Design Museum in Berlin. Skira Editore, Milan, 2002
« Il design Cartier visto da Ettore Sottsass » at Palazzo Reale, Milan, Skira Editore, Milan, 2002
"Cartier design viewed by Ettore Sottsass". Skira Editore, Milan, 2002
"Cartier design viewed by Ettore Sottsass". Skira Editore, Kyoto, 2004
"Cartier design viewed by Ettore Sottsass". Skira Editore, Houston, 2004

VON HABSBURG, Geza.
Exhibition catalogue « Fabergé-Cartier. Rivalen am Zarenhof » at Kunsthalle der Hypo-Kulturstiftung, Munich.

Hirmer Verlag, Munich 2003.

COLOGNI, Franco, FORNAS, Bernard, XING Xiaozhou, CHEN, Xiejun, BAO, Yanli.
Exhibition catalogue "The Art of Cartier" at Shanghai Museum.
The Shanghai Museum, Shanghai 2004.

FORNAS, Bernard, LEE Chor Lin
Exhibition catalogue "The Art of Cartier" at National Museum of Singapore.
The National Museum of Singapore. 2006.

FORNAS, Bernard, VILAR, Emilio Rui, CASTEL-BRANCO PEREIRA, João, VASSALO E SILVA, Nuno, PASSOS LEITE, Maria Fernanda, RAINERO, Pierre, RUDOE, Judy, REMY, Côme, COUDERT, Thierry.
Cartier, 1899-1949, The journey of a style at the Calouste Gulbenkian Foundation
Milan: Skira Editore 2007
Cartier, 1899-1949, Le parcours d'un style
Milan: Skira Editore 2007
Cartier, 1899-1949, O percurso de um estilo
Milan: Skira Editore, 2007
Cartier, Cartier 1899-1949 : El recorrido por un estilo
Milan: Skira Editore, 2007

FORNAS, Bernard, GAGARINA, Elena, ALIAGA, Michel, CHAILLE, François, MARIN, Sophie, MILHAUD, Pascale, PESHEKHONOVA, Larissa, RAINERO, Pierre
Exhibition catalogue « Cartier, innovation through the 20th century » at the Moscow Kremlin Museums.
English version and Russian version.
Editions Flammarion 2007.

FORNAS, Bernard, KIM, Youn Soo, LIU, Jienne, RAINERO, Pierre, MILHAUD, Pascale,
Exhibition catalogue „The Art of Cartier" at the National Museum of Art, Deoksugung Seoul
English version and Korean version.
National Museum of Contemporary Art, Korea/Cartier - 2008.

NIKKEI & TNM, FAURE, Philippe, FORNAS, Bernard, TOKUJIN, Yoshioka, RAINERO, Pierre, LAURENT, Mathilde, MAEDA, Mari
Exhibition catalogue "Story of… , Memories of Cartier creations" at the Tokyo National Museum
English and Japanese version.
Nikkei Inc.

3 – Privately printed publications

Cartier New York
Editions Assouline, Paris, 2001

Cartier London
Editions Assouline, Paris, 2002

Cartier et la Russie
Editions Assouline, Paris, 2003

Cartier 13 rue de la Paix
Editions Assouline, Paris, 2005

后 记

　　"卡地亚珍宝艺术展"图录是为配合卡地亚世纪珍宝特展而编辑出版，收录了此次展览的全部展品 346 件。为了促成世界享有盛誉的时尚品牌卡地亚珍宝在故宫展出，故宫博物院和卡地亚公司都做出了积极努力。在筹备"卡地亚珍宝艺术展"的过程中，我们专程赴法国、瑞士卡地亚公司遴选展品，得到卡地亚公司的大力支持和积极配合。具有一百六十多年历史的卡地亚吸纳、学习世界各民族艺术，始终保持创新意识，追求精湛、细腻的工艺，用充满文化内涵的设计诠释时尚、优雅、高贵和奢华。当数百件精美珠宝饰品从我们眼前经过的时候，我们深深地被卡地亚作品所拥有的广泛而深刻的文化内涵所震撼，被不同主题作品的恒久魅力所折服。

　　我们希望奉献给观众的是一次时尚装饰盛宴。在此，我们对展览双方所有为此次展览和图录编辑给予支持和帮助的同行、同事、朋友表示衷心的感谢。由于我们对时尚、潮流的理解和卡地亚的历史研究有限，展览和图录难免有错误和不足之处，欢迎各界朋友批评指正。

编委会

2009 年 9 月

For the purpose of facilitating the exhibition titled "Cartier Treasures-King of Jewellers, Jewellers to Kings" at the Exhibition Hall in the Meridian Gate, we specially compiled this album. It includes not only all the 346 exhibits, but also research articles written by experts from the Palace Museum and Cartier for the purpose of enriching the content of the album.

Both the Palace Museum and Cartier have made proactive efforts to organize this exhibition. We made special trips to France and Switzerland to select exhibits during the preparatory period, which were greatly supported by Cartier. Cartier, with a history of over 160 years, has been learning from and assimilating different cultures, persisting in innovation and pursuing exquisite techniques. It has been interpreting fashion, elegance, nobleness and luxury with its works' rich cultural meanings. When we saw hundreds of splendid treasures, we were deeply touched by the wide and profound cultural implications and the eternal charms of the theme designs.

We wish that all our efforts would finally lead to a visual feast for spectators.

We hereby express our sincere appreciation for our colleagues who have supported and assisted the exhibition and the album, and for all people who have contributed. It is your efforts that have made the exhibition and album possible, and we are glad about that.

Due to our limited knowledge about fashion, trend and the history of Cartier, there may be inevitably mistakes and deficiencies on the exhibition and album. We sincerely welcome criticisms from all of you.

Editorial Board

September, 2009

编辑出版委员会

EDITORIAL COMMITTEE

图书在版编目（ＣＩＰ）数据

卡地亚珍宝艺术／故宫博物院编 .—北京：紫禁城出版社，2009.10 重印

ISBN 978-7-80047-859-8

Ⅰ.卡… Ⅱ.故… Ⅲ.首饰—法国—图集 Ⅳ. TS934.3-64

中国版本图书馆 CIP 数据核字（2009）第 150764 号

本书图版说明、年表、参考书目均由卡地亚提供；

图版由尼克·威尔士提供；

全部中文文稿及图片均由故宫博物院整理、加工。

The captions, chronology and bibliography are from Cartier.

The images in the plates are from Nick Welsh, Cartier Collection © Cartier

The Chinese texts and images are edited by the Palace Museum.

卡地亚珍宝艺术

故宫博物院编

主　　编：宋海洋

英文审校：姜斐德　张　彦

责任编辑：万　钧　方　妍

装帧设计：李　猛　燕军君

出版发行：紫禁城出版社

　　　　　地址：北京东城区景山前街 4 号　邮编：100009
　　　　　电话：010-85007808　010-85007816　传真：010-65129479
　　　　　邮箱：gugongwenhua @ yahoo.cn

制版印刷：北京雅昌彩色印刷有限公司

开　　本：889×1194 毫米　1/16

印　　张：31

版　　次：2009 年 9 月第 1 版
　　　　　2009 年 10 月第 2 次印刷

印　　数：5,501~7,500 册

书　　号：ISBN 978-7-80047-859-8

定　　价：480.00 元